THE WORLD OF MINERALS

THE WORLD OF MINERALS

Vincenzo de Michele
Curator of the Department of Mineralogy
at the Museum of Natural History, Milan

With a foreword by
G. F. Claringbull PhD, FInstP, FGS

ORBIS PUBLISHING · LONDON

The author acknowledges the help and co-operation of
Professor Cesare Conci, Director of the Museo Civico
di Storia Naturale in Milan. All the minerals
photographed are from this collection.

The line drawings are by Emilio Chiusa

Title page: gold panning (Mansell Collection)
Endpapers: malachite (British Museum, Natural History)

© Istituto Geografico de Agostini, Novara 1971
English edition © Orbis Publishing Limited, London 1972
Revised English edition © Orbis Publishing Limited, London 1976
Printed in Italy by IGDA, Officine Grafiche, Novara
ISBN 0 85613 203 9

For many years as Keeper of Mineralogy I had charge of the collections of minerals in the British Museum (Natural History), one of the finest in the world, and worked on their identification and crystal structures. Despite this familiarity, however, I am still fascinated and thrilled by their intrinsic beauty of colour, form and pattern. To the many who are not privileged even to have easy access to a fine mineral collection – of which incidentally there is only a small number throughout the world – the excellent illustrations in 'The World of Minerals' will be a compensation, and will bring some of the beautiful specimens into their homes to be admired at leisure. In the associated text much interesting and useful information is provided about the wide range of the illustrations. Not only is the chemical and physical nature of many mineral species described, but a wide range of other facts about their origin, crystal form and mode of origin are given.

The publishers are to be congratulated in having brought together such a fine collection of coloured illustrations of minerals, and I commend this book to all who wish to become acquainted with the world of minerals and discover the extraordinary variety of its manifestations.

G. F. Claringbull PhD, FInstP, FGS
Director of the British Museum (Natural History)

Contents

Index of minerals

The roman numbers refer to the page on which a mineral
is described; italic numbers indicate the plates

THE MINERAL KINGDOM

All physical bodies in Nature can be divided into three primary groups called the animal, vegetable, and mineral kingdoms; everything that cannot be assigned to the first two groups belongs to the last. So the mineral kingdom is made up of all lifeless matter contained in our planet, the universe around it and, by extension, all the organic and inorganic products of living creatures, together with the various results of interactions and activities of elementary particles, light, electrical phenomena, etc.

Within this vast array of heterogeneous substances and phenomena of every kind are found *minerals*, which are the result of specific chemical-physical actions on, and changes in, matter, and can be regarded as the smallest unit in the huge geological picture presented to us by the history of the Earth's crust, because minerals, when mixed together, are the materials of which rocks are made. However we should not confuse *minerals* with the *mineral kingdom*, or the pure *mineral* with *mineral substances*.

But what, then, is a mineral?

No doubt the best way to decide upon whether a substance is a mineral or not, is to identify the general properties which are in themselves sufficient to differentiate minerals from other inanimate substances.

Two of these properties are immediately apparent: *for a substance to be classified as a mineral it must be homogeneous, and solid under normal temperature and pressure conditions.* By 'homogeneous' we mean a substance which contains the same constituent elements in all its parts; so if we detach two pieces from two different parts of a block of pyrite and analyze them, they should have an identical chemical composition, and the same should happen even if the two pieces are collected from two different worlds. If on the other hand we should carry out the same tests on other substances such as bauxite, porphyry, granite, or sand, we would notice that their composition varies considerably from one side of the sample to the other and from one sample to another of the same type.

Therefore we must exclude from the ranks of true minerals all heterogeneous masses such as rocks, if they are composed of more than just one mineral.

We know from elementary physics that true 'solid' bodies are those that possess an internal structure of atoms linked together in a perfectly ordered lattice, in which they appear in a specific three-dimensional arrangement. Such a substance is called 'crystalline' because 'crystals' are the naturally formed geometric shapes resulting from this structure.

Among the other physical properties deriving from this internal structure of matter, is the significant one concerning the melting point of substances. True solid bodies can be recognized by the fact that when the temperature is increased sufficiently they suddenly pass from a solid to a liquid state, and the temperature at which this occurs is constant for each specific body.

Other apparently solid bodies are in reality simply hard ones. So glass, resin, and colloids, for example, do not have a set melting point but change from a state of hardness to one of plasticity, then softness, and finally fluidity.

Substances behaving in this manner are not minerals; therefore we can exclude gases, liquids, and those we have mentioned above. And while it has traditionally been accepted that mercury, opal, and chrysocolla, among others, should be grouped with the minerals, strictly speaking they should not be considered as such.

In addition, *minerals should be natural substances in the Earth's crust* and so must be the result of natural phenomena and not the artificial intervention of man; therefore they are the faithful witnesses of the history of our planet and are invaluable in enabling us to reconstruct, through the study of rocks, the complex transformations in the Earth's crust.

So we cannot include among the minerals all those inorganic salts produced by the chemical industry, even if they have analogous compositions and the physical properties of minerals; it would be incorrect to consider the salt obtained from salt-works as a mineral, though it is chemically identical to rock-salt. The latter, in fact, is the result of one particular type of geological process in the history of the Earth, and has its logical place in the succession of changes undergone by the Earth's crust, in which it has an essential and illuminating part. The artificial product cannot have any such significance.

An interesting point is that the minerals present in meteorites which are, of course, orbiting bodies in the solar system that have fallen to our Earth, are considered true minerals even though they do not originate from our planet. This makes sense if we expand the concept of 'natural substance in the Earth's crust' to include the natural elements of other celestial bodies; moreover, it is reasonable to consider meteorites which have fallen on our Earth as forming an integral part of it.

To go further into the statement we made at the beginning, namely that all living organisms are excluded from the mineral kingdom, we can state that *minerals originate from inorganic processes*, and therefore the products of plant secretions and the accumulation of organisms cannot be included in our category. So amber, resin, hydrocarbons, bitumens, etc, are not minerals.

To complete the picture, it is essential that minerals should be *characterized by a uniform chemical composi-* *tion, or vary only within fixed limits, and that they should present well-defined physical properties*. This means that a mineral, whatever its appearance, should always have the same composition, expressed in a chemical formula which has been established as the result of the analysis of a large number of examples. For example, pyrite, an isometric sulphide of iron, FeS_2, has a constant 46.55 per cent iron and 53.45 per cent sulphur; galena, a sulphide of lead, PbS, contains 86.60 per cent lead and 13.40 per cent sulphur. As already mentioned, the chemical elements are arranged in a different atomic structure for each mineral; this structure determines the crystalline nature which is the basis of the physical properties, and is constant for each mineral, such as the geometric shape, specific gravity, refractive index, melting point, etc.

In conclusion, let us recapitulate by defining minerals as being *homogeneous solid bodies, occurring naturally in the Earth's crust, formed by inorganic processes, and characterized by a number of constant physical properties.*

Approximately 1,900 different types of minerals are known to date.

Classification of minerals

The definition of the mineral state as outlined in the previous chapter has another extremely important aspect, which is that while enabling us to differentiate these substances from the rest of the non-biological forms in the mineral kingdom, it also provides the means by which minerals can be grouped in classes or families according to the degree of their similarity to one another.

The characteristics on the basis of which this grouping can be effected are the chemical composition and crystalline structure, the first obtained by chemical analysis and the second by X-ray techniques.

As chemical compounds, minerals have this peculiarity: while they are made up of a great many electropositive elements (cations)*, such as iron, magnesium, zinc, copper, lead, sodium, and potassium, they have very few electronegative elements (anions), such as sulphur, fluorine, and oxygen. Therefore it is much simpler to group minerals according to their anions, which thus become the unifying factor for substances that at first glance appear completely different.

*Positive ions (cations) are atoms lacking one or more electrons; negative ions (anions) are atoms having an excess of one or more electrons.

In addition, as in the other natural sciences, botany and zoology, the classification of minerals also has the simplest entities as its starting point and then progresses to the more complicated. The anion is the principal factor in this increasing degree of complexity.

Therefore, in accordance with this rule, the first mineral class contains simple types containing only native elements such as copper, silver, and so on. After this we go on to more complex classes containing elements combined with sulphur (*sulphides*), with fluorine, chlorine, bromine, or iodine (*halides*), or with oxygen (*oxides*). At this point we arrive at a second increase in complexity: the cationic element may combine with more than one anion, for example with sulphur and oxygen to form a *sulphate*; with carbon and oxygen to form *carbonates*, with phosphorus and oxygen to form *phosphates*; and finally with silicon and oxygen to form *silicates*. Within each class there is a further redistribution according to an increasing range of complexity. In the case of the silicates there is a progression from the simplest nesosilicates to the most complex tektosilicates.

The classification of minerals in this book is laid out as follows:

Class I   *Native elements*
Class II   *Sulphides, selenides, tellurides, arsenides, antimonides, bismuthides*
Class III   *Halides*, namely, *fluorides, chlorides, bromides, iodides*
Class IV   *Oxides* and *hydroxides*
Class V   *Carbonates, borates, nitrates*
Class VI   *Sulphates, tellurates* and *chromates, molybdates,* and *tungstates*
Class VII   *Phosphates, arsenates, vanadates*

Class VIII   *Silicates*, with the sub-groups: nesosilicates, sorosilicates, cyclosilicates, inosilicates, phyllosilicates, tektosilicates

The name of each mineral is usually followed by its chemical formula, then its crystal form (habit), specific gravity (sg), and degree of hardness. In addition to a general description of their appearance, there is also a list of some of the more important localities where these minerals can be found.

Glossary of terms

Acicular. Needle-like
Adamantine. A kind of lustre, like diamond
Amorphous. Formless; not crystalline
Bacciliform. Rodshaped
Basic. Loosely, alkaline; with a low silica content
Bi-sphenoid. See Sphenoid
Botyroidal. Having spherical bumps on the surface
Brecciated. Composed of fragments
Calcareous. Chalky
Capillary. With fine-bored tubes
Cleavage. Tendency to split along flat surfaces
Colloidal. Finely distributed in a medium
Conchoidal. Shell-like, rippled
Concretionary. A mass accumulated around a point
Diabase. Rock similar to basalt
Druse. A rock lined with crystals
Ductile. Capable of extension into threads
Effervescence. Formation of bubbles in solution
Efflorescence. A powdery crust on the surface
Felted. Matted together with fibres
Ferric, ferrous. The two cationic forms of iron
Filiform. Threadlike
Fracture. The broken surface of a stone
Globular. Roughly spherical
Habit. The crystal form of a mineral
Hardness. Resistance to scratching
Hexagonal system. A system with three axes meeting at an angle of 60°, and the fourth perpendicular
Icositetrahedron. A crystal with twenty-four faces
Isometric system (cubic system). A system with three axes at right angles to each other
Lacustrine. Formed in lakes
Lamellar. Composed of thin layers
Lenticular. Shaped like a biconvex lens
Leucitotetrahedron. A type of icositetrahedron
Limonitic. Containing brown iron ore
Lustre. The brilliance, or shine, of a mineral
Magmatic. Formed of molten rock
Mammillary, mammellar. Shaped like a breast
Metamorphosis. A change from one form to another
Modification. A change due to the environment
Monoclinic system. A system with three unequal axes; two intersecting, the third perpendicular

Orthorhombic system. A system with three unequal axes at right angles
Patina. A film formed on the surface of a mineral
Pegmatite. A coarse crystalline granite
Peridotite. A type of coarse igneous rock
Pinacoid. A type of crystal face
Pleochroism. Colour change under polarized light
Polymorphs. Different crystal forms of one compound
Polysynthetic twins. Many thin twinned lamellae
Prism (atic). With several faces parallel to an axis
Pseudo . . . Simulating some crystalline form
Pseudomorph. An altered crystal, retaining its form
Radical. A charged chemical group, eg NO_3^-, NH_4^+
Rare earths. A group of uncommon, similar elements
Reniform. Kidney-shaped
Reticulate. Of a net-like structure
Rhomb . . . With rhomboid surfaces
Rosette. A radiating, concentric appearance
Saccharoidal. Like loaf sugar in texture
Scalenohedral. With sides that are scalene triangles
Schist. A rock with layers of different minerals
Selliform. Saddle-shaped
Serpentinites. A green type of rock formation
sg. Specific gravity; weight compared with water
Silicified. Impregnated with silica
Spathic. Like spar
Sphenoid. Wedge-shaped, with four triangular faces
Stalactite. A mineral column hanging downward
Stalagmite. A mineral column growing upward
Syenite. A type of coarse-grained rock
System. A classification of crystals, defining them according to the axes along which they are formed
Tabular. Flattened horizontally, like a table
Tetragonal system. A system with three axes at right angles, the two lateral ones equal
Triclinic system. A system with three unequal axes none at right angles. It has no planes of symmetry
Trigonal system (rhombohedral system). Like the hexagonal system, but with fewer planes of symmetry
Trisoctahedron. With twenty-four faces, 8×3
Tuff. A rock composed of volcanic fragments
Twins. Two or more similar crystals, joined in a certain way; eg at an angle, or slightly rotated
Vitreous. Like glass

THE MINERALS

Class I: native elements

Chemical elements found in Nature in a free, uncombined state are known as *native* elements. There are scarcely more than a dozen of them and some, such as gold, silver, and copper, have played an important part in the history of man, chiefly because they were the first metals known and used in prehistoric times.

Copper, Cu: isometric system; sg 8.95; hardness 2.5–3.

Owing to its relative abundance and high malleability, it was the first metal to be used for the making of tools in prehistoric times (the Copper Age, which was followed by the Bronze Age when copper was mixed with tin). Its characteristic copper-red colour is often hidden by green patinas of malachite formed by oxidation. It is generally found in dendritic, filiform, or massive formations; it frequently has dodecahedral or cubic crystals.

The largest deposits have been found in the Precambrian Shield area of central Canada and the northern United States (especially northern Michigan), the Rocky and Andes mountains, and central Africa. Excellent examples occur at Bogoslovsk in the Urals and Cornwall in England.

Mercury, arsenic, antimony

These native elements are not very common and have little practical use.

Mercury, Hg: liquid (at −39°C it solidifies and becomes trigonal); sg 13.59. It is found in liquid drops in cinnabar at Idrija (Yugoslavia), some mines on Mount Amiata, Tuscany (Italy), and Almadén (Spain). In the United States it is found in California, Alaska, Oregon, Texas, and Arkansas.

Arsenic, As: trigonal system; sg 5.63; hardness 3.5. It is found in masses of a concentric structure, coloured dark gray or brownish black, at Freiberg and Andreasberg in Germany, and also in the United States (Arizona).

Antimony, Sb: trigonal system; sg 6.61; hardness 3–3.5. Normally in compact masses coloured metallic gray with a metallic lustre. Bolivia contains the greatest

deposits, although there are significant quantities in the western United States and Canada.

Platinum, Pt: isometric system; sg 21.46; hardness 4–4.5.

When it was first discovered in Colombia in the eighteenth century, it was mistaken for silver which in Spanish is called *plata*; hence the name platinum. Coloured silver white, it is always found in igneous rocks rich in magnesium and poor in silica, such as gabbro and peridotite. It is normally found in granules and nuggets in sand formed by erosion of platinum-bearing rocks; this is the origin of the

Filiform silver. Nicola Secci, Sàrrabus, Sardinia (Italy)

4

Dendritic copper formation. Houghton (Michigan)

Plates of gold on quartzite. California

famous Perm deposits in the Urals, discovered in 1822. It also occurs in Australia, New Zealand, Alaska, and at Sudbury, Ontario.

Graphite, C: hexagonal system; sg 2.09–2.21; hardness 1–2.

Only occasionally in crystals, it is usually found in compact masses, coloured black with a semi-metallic lustre, and greasy to the touch. It is found in igneous rocks. Large deposits are found in Siberia, near Irkutsk, and in the Austrian Alps. The chief deposits in the United States are in the Adirondack region of New York. Its high melting point makes it suitable for use as a refractory, a dry lubricant, and for electrodes and atomic piles.

Silver, Ag: isometric system; sg 10–11; hardness 2.5–3.

Native silver has no great importance from the mining point of view due to its comparative rarity; it is found mainly in the upper layers of lead-silver deposits.

It has a characteristic silver-white colour but tends to blacken when exposed to air; ductile and malleable, it has a reticulate, arborescent or convoluted form; occasionally it can form cubic or octahedral crystals of small dimensions.

The most famous deposits of silver are in the North American continent, in various regions of Mexico, the western United States (especially the Comstock Lode in Nevada), and in the Canadian Shield (Sudbury, Ontario).

Gold, Au: isometric system; sg 19.3; hardness 2.5–3.

It would be superfluous to dwell on the great interest this mineral has always aroused. It is usually present in reticulate, dendritic, or filiform aggregates, while octahedral or cubic crystals are rather rare; it is usually found in association with pyrite, chalcopyrite, and arsenopyrite in quartz veins.

The typical colour is yellow, but a different yellow from that of pyrite or chalcopyrite, and sometimes it tends towards white due to a silver content. When found in stream deposits, the gold is in small flat particles or masses (nuggets) of varying sizes. It is very ductile and malleable.

The most important deposits in the world are centred on the Witwatersrand in the Transvaal (South Africa), and indeed South Africa currently produces well over half the world's supply. The other leading suppliers are, in order, the USSR, Canada, and the USA (South Dakota, Nevada, Utah, Arizona, and California). The major locality in Europe is Rosia-Montana in Romania, where gold is found with sylvanite and nagyagite.

Diamond, C: isometric system; sg 3.50; hardness 10.

Though diamond and graphite have the same chemical composition – pure carbon – they could not look more different. Theirs is the most notable case of diversity in structure and properties shown in two modifications of the same chemical element. While graphite is black, opaque, and soft, the diamond is lustrous, transparent, colourless, yellow or green, and very hard; graphite is hexagonal and lamellar, while diamond is isometric and forms crystals of an octahedral habit; graphite is flexible and yielding, diamond is brittle and hard. This can be explained by the atomic structure of the two minerals: in graphite the atoms are arranged in layers, while in the diamond they are much more compact.

The name diamond is a corruption of a Greek word meaning 'invincible', so called because of its great hardness. Its use as a precious stone goes back to very ancient times when it was found in the alluvial deposits of Borneo and India. It was not till the eighteenth and nineteenth centuries that other deposits were found in Brazil (Minas Gerais) and South Africa (Transvaal, along the Orange River, and especially near Kimberley); diamonds from the Kimberley mine are embedded in an ultrabasic rock called *kimberlite*, which also contains pyrope and diopside. Other famous mines are Dutoitspan, Bultfontein, De Beer's, and Premier; in 1905 the biggest diamond in the world (weighing 3,106 carats, equal to 621 grams) was found at the Premier Mine; named the Cullinan diamond, it was presented to the King of England.

Sulphur, S: orthorhombic system; sg 2.07; hardness 1.5–2.5.

This mineral has an attractive appearance and frequently has good dipyramidal crystals which are bright yellow, brown yellow in the bituminous varieties, or reddish yellow, with a resinous translucent lustre.

It can occur in microcrystalline crusts or compact masses deposited by volcanic gases (eg H_2S) as at Pozzuoli, Naples (Italy); Milos in Greece; Japan, and in the craters of active volcanoes. Small crystals are formed by the alteration of some sulphides such as galena, antimonite, and pyrite. Finally, it can have a sedimentary origin as in the big Italian and American deposits (Texas and Louisiana). Good crystals associated with bituminous deposits have been found in the sulphur mines of Romagna, Italy, in association with gypsum and celestite. In Sicily, on the other hand, particularly fine crystals occur and they are associated with clear gypsum, pseudo-hexagonal aragonite, clear, white, or blue celestite, and calcite.

Right: Groups of sulphur crystals partially covered with calcite. Casteltermini, Agrigento (Italy)

Left: Diamond crystal in kimberlite rock. Kimberley (South Africa)

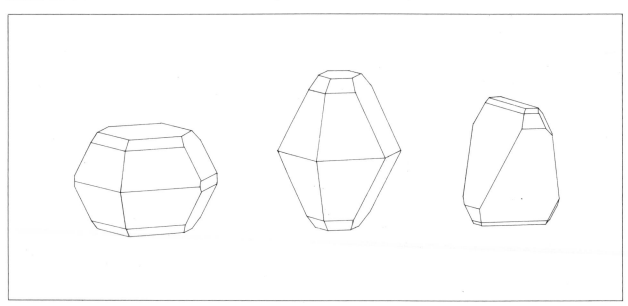

Bipyramidal orthorhombic sulphur. Left: Two individuals from Tuscany. Right: One individual from Sicily

Class II: sulphides

In addition to compounds of metals with sulphur, this class includes compounds with selenium (*selenides*) tellurium (*tellurides*), arsenic, antimony, and bismuth (*arsenides*, *antimonides*, *bismuthides*). Most of these minerals can be recognized by their metallic lustre, and many are used for commercial or industrial purposes.

Chalcocite, covellite, bornite

Chalcocite, Cu_2S, orthorhombic, rarely forms crystals but more often microcrystalline compact masses with a lead-gray colour and conchoidal fracture. An important copper mineral, it is frequently accompanied by bornite and covellite. Fine pseudo-hexagonal crystals come from Redruth in Cornwall and Messina in South Africa. Exceptional crystals come from Bristol, Connecticut, and it is also found at Butte, Montana, and the Copper River district of Alaska.

A much rarer mineral is *covellite*, CuS, hexagonal, in lamellar metallic-blue crystals, from Calabona in Sardinia, and Butte, Montana.

Another useful copper sulphide is *bornite*, Cu_5FeS_4, isometric, and rarely in separate crystals. It is usually found in compact microcrystalline masses, of a bronze colour on a fresh fracture but quickly tarnishing to

Far left: Lamellar covellite. Calabona, Sardinia (Italy)

Left: Chalcocite. Cornwall (England)

Right: Sphalerite in its black ferriferous variety called marmatite, with pyrite and quartz. Rodna Veche (Romania)

purple and violet (hence called *erubescite or peacock ore*). It can be obtained in crystal form at Pragratten in the Tyrol (Austria), Redruth in Cornwall (England), and Bristol, Connecticut. It also occurs at Butte, Montana, and in the eastern part of Quebec.

Argentite

Argentite, Ag$_2$S, isometric, is always associated with native silver. It frequently occurs in cubic crystals or compact masses with a lead-gray colour and metallic lustre that tarnishes on exposure to air. Important deposits from which the best examples come, are the German ones of Freiberg, Schneeberg, and Marienberg, and the Czechoslovakian ones of Kremnica, Banská Štiavnica, and Příbram. In the United States it has been an important ore mineral in Nevada, notably at the Comstock Lode and at Tonopah. It is also found in the silver districts of Colorado, and at Butte, Montana.

Sphalerite, ZnS: isometric system; sg 3.9–4.1; hardness 3.5–4.

Also known as *zinc blende,* this mineral frequently forms splendid crystals with a tetrahedral habit, the colour of which varies from yellow to brown, red, or black, with a resinous or adamantine lustre, transparent or translucent. The ferrous variety is called *marmatite.*

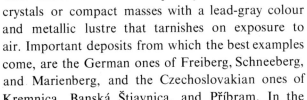

Above: Red sphalerite. (Wales)

Left: Pseudo-tetrahedral chalcopyrite crystals. Ani, Hugo (Japan)

10

A very important zinc mineral, it is usually found in massive, cleavable, or granular masses forming veins in limestone, or concretionary deposits (*schalenblende*).

Almost always associated with galena, it forms enormous deposits in Oklahoma, Kansas, and Missouri, while important deposits are also exploited at Touissite and Bou-Beker in Morocco, and Broken Hill, Zambia. Beautiful ruby-red crystals come from the above three American states, especially from Joplin, Missouri; brown-red ones are obtained at Binnatal in Switzerland and at Picos de Europa near Santander in Spain. Very lustrous black crystals of marmatite blende are found in the Bottino mines near Seravezza in the Apuanian Alps (Italy); red blende crystals are mined at Seddas Moddizzis near Iglesias in Sardinia.

Wurtzite, greenockite

Wurtzite is a hexagonal modification of zinc sulphide, coloured brownish black, and less common than blende.

Usually found in association with blende, is the cadmium sulphide, *greenockite*, CdS, in greenish-yellow coatings.

Chalcopyrite, $CuFeS_2$: tetragonal system; sg 4.1; hardness 3.5–4.

It can occasionally be found in tetrahedral crystals (which are really bi-sphenoid) although usually it forms compact masses with a brass-yellow colour and a metallic lustre enlivened by a reddish and bluish iridescence caused by alteration. It can be confused with pyrite which has a blackish streak, whereas chalcopyrite gives a greenish streak. It is the most widely found copper ore and there are rich deposits in the United States (for example Arizona and Utah). Fine crystals come from Pennsylvania and Colorado, and very large crystals and massive pyrite is found at Ellenville, Ulster County, New York. This very widespread mineral is to be found in most countries.

Tetrahedrite, Cu_3SbS_3: isometric system; sg 4.62; hardness 3.5–4.5.

Tennantite, Cu_3AsS_3: isometric system; sg 4.62; hardness 3.5–4.5.

As the name indicates, the first of these minerals may form tetrahedral crystals with a gray or black colour and metallic lustre. More often, however, it is compact and granular. *Tennantite*, in which antimony (Sb) is replaced by arsenic (As), has much the same appearance, but it is much rarer. Both minerals usually contain some iron; when tetrahedrite contains silver it is called *freibergite*, and when it contains mercury, *schwatzite*; *frigidite* is the name given to a nickeliferous variety existing in the Apuanian Mountains; *binnite* is a silver tennantite found in dolomite at Binnatal, Switzerland. Good crystals

Below: Tetrahedrite covered with chalcopyrite. Cornwall (England)

occur at Freiberg in Saxony, Cavnic in Romania, Schwaz in the Tyrol (Austria), and Cornwall, England. In the United States good tetrahedrite crystals are found near Central City, Colorado, and in many other localities.

Pyrrhotite, FeS: hexagonal system; sg 4.58–4.65; hardness 3.5–4.5.

This mineral, the name of which comes from the Greek word meaning 'reddish', usually occurs in compact, very flattened granular masses, or lamellar crystals, or rose-shaped aggregates of a bronze colour. There is a very large nickeliferous pyrrhotite deposit at Sudbury, Ontario. Fine crystals are found at Trepča in Yugoslavia and Chiuzbaia in Romania. In the United States it comes from Standish, Maine, and from localities in the states of New York, Pennsylvania, and Tennessee.

Niccolite, millerite

Niccolite, NiAs, is a hexagonal arsenide of nickel, compact, coloured bronze with a touch of pink, and found at Cobalt, Ontario; Franklin, New Jersey; and Silver Cliff, Colorado.

Millerite NiS, belongs to the trigonal system and is found in thin acicular or capillary brass-yellow crystals with a metallic lustre, mainly at Siegerland in Germany and in Bohemia (Czechoslovakia). It is found at St Louis, Missouri, Keokuk, Iowa, and other localities in the United States.

Galena, PbS: isometric system; sg 7.58; hardness 2.5.

This very common mineral has the highest lead content of the lead minerals and forms compact or granular masses, though not infrequently it can have quite large cubic or octahedral crystals; the colour is lead gray and freshly cut surfaces show a very bright metallic lustre; the cleavage is cubic. It can have a high silver content which is why it is considered a useful ore of this metal. Very well known are the huge cubic crystals on pink dolomite mined at the famous Three States deposit on the border between Missouri, Kansas, and Oklahoma. It occurs at Mineral Point and Schullsburg in Illinois, at Galena in eastern Iowa, and it is found with crystals of calcite and chalcopyrite at Rossie, St Lawrence County, New York. It is found in many other localities, among them the celebrated Bleiberg mines in Carinthia (Austria), Freiberg in Saxony (Germany), Alson Moor in Cumberland (England), and Broken Hill in New South Wales (Australia).

Cinnabar, HgS: trigonal system; sg 8.09; hardness 2.5.

It is the most important ore of mercury, usually accompanied by antimonite, realgar, and rare drops of native mercury.

The vermilion red colour, tending to brown red, is a characteristic of this mineral's compact and

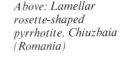

Above: Lamellar rosette-shaped pyrrhotite. Chiuzbaia (Romania)

Left: Acicular millerite crystals. Westwald, Nassau (Germany)

granular masses; sometimes it also forms crystals, as at Hunan in China, Almadén in Spain, and Ripa in Tuscany, Italy. There are very famous deposits at Idrija in Yugoslavia and Mount Amiata in Tuscany; very old mines are at Vallata (Belluno), Margno (Como), and Levigliani (Lucca), all in Italy. In the United States the most important deposits are in California.

Metacinnabarite

Metacinnabarite, HgS, has the same chemical composition as cinnabar, but it is isometric. It is a rare modification of a sulphide of mercury and forms little shiny black masses or small pseudo-octahedral tetrahedral crystals. It is found at numerous localities in California.

Stibnite, Sb_2S_3: orthorhombic system; sg 4.63; hardness 2.

This is a very distinctive-looking mineral because it has beautifully formed prismatic crystals, striated along their length and coloured steel gray, with a bright metallic lustre which can sometimes be dulled by incipient alteration. The cleavage is prismatic. Groups of crystals as much as 50 centimetres (20 inches) long have been obtained at the Ichinokawa mines on the Island of Shikoku in Japan. Other notable

Cuboctahedral galena. Freiberg, Saxony (Germany)

Concretionary cinnabar. Mt Amiata, Tuscany (Italy)

localities are at Wolfsberg, Harz Mountains (Germany), and Kremnica in Czechoslovakia. Stibnite occurs infrequently in the United States; it is however, found at Hollister, Benito County, California. The compact variety is found at Gerrei in Sardinia.

Bismuthinite, Bi_2S_3: orthorhombic system; sg 6.78; hardness 2.

Similar in appearance and colour but somewhat rarer and only occasionally found in single crystals, is *bismuthinite*, a sulphide found in Bolivia, Cornwall (England), and Saxony (Germany). In Italy it is found at Brosso near Ivrea (Piedmont) in association with siderite and scaly hematite, and at Alfenza, near Crodo in Val d'Ossola (Piedmont). In the United States it has been found in Connecticut, Pennsylvania, Montana, and Utah.

Pyrite, FeS_2: isometric system; sg 5.01; hardness 6–6.5.

Commonly known as 'fool's gold', this is the most abundant and widely diffused sulphide on the Earth's crust. It can be in compact or granular masses, or in many different types of crystals: striated cubes ('triglyph'), pentagonal dodecahedrons (also called pyritohedrons), octahedrons, etc. The colour is pale brass yellow, the lustre metallic; it can be distinguished from chalcopyrite because it has a black instead of a greenish streak. It can contain a certain percentage of gold,

Above: Pyrite in pentagonal dodecahedra. Rio Marina, Elba (Italy)

Above right: Marcasite. Cornwall (England)

Pseudo-orthorhombic prisms of arsenopyrite. Trepča (Yugoslavia)

15

16

when it is known as 'auriferous pyrite'.

There are numerous pyrite deposits; among the largest is the cupriferous pyrite vein at Rio Tinto (Huelva, Spain). Fine specimens come from southwest England, and from many other countries.

Marcasite, hauerite, cobaltite, ullmannite

Marcasite is an orthorhombic modification of iron sulphide which forms aggregates of tabular crystals

or spherical nodules with a fibrous radiating structure, coloured brass yellow but often altered to brown by oxidation. The rounded, concretionary forms which can sometimes be confused with meteorites, are commonly found in clay and sometimes marly limestone.

There is a rare isometric manganese sulphide called *hauerite*, MnS_2, found in brown opaque octahedra in clay at Destricella Mine near Raddusa (Enna) in Italy.

A brief mention should be made of the two isometric sulphides: *cobaltite*, CoAsS, in cubes or pyritohedron crystals, slightly pinkish white, from Tunaberg (Sweden) and Skutterud (Norway); and *ullmannite*, NiSbS, which has gray-white cubic crystals and is found on Mount Narba near Sàrrabus, Sardinia.

Arsenopyrite, FeAsS: orthorhombic system; sg 6.07; hardness 5.5–6.

It generally shows prismatic pseudo-rhombic, silver-white crystals with a not very vivid metallic lustre. It can contain some gold, and is the most abundant and widely occurring arsenic mineral. Good crystals come from Freiberg in Saxony (Germany), Cornwall (England), and Trepča (Yugoslavia); and from Franconia, New Hampshire; Franklin, New Jersey; Emery, Montana; and Leadville, Colorado.

Molybdenite, MoS_2: hexagonal system; sg 4.62–4.73; hardness 1–1.5.

This is the best-known molybdenum mineral and usually forms flexible, non-elastic, leaf-like plates of a lead-gray colour. It occurs in granites, porphyries, and pegmatites. Good examples are found at Deepwater and Kingsgate in New South Wales, Australia, and it has been found at Blue Hill Bay, Maine, Haddam, Connecticut, and Frankford, Pennsylvania among other localities in the United States.

Skutterudite

Skutterudite, or *smaltite*, is an isometric cobalt arsenide with a chemical composition of $CoAs_3$, abundant in Germany and Morocco. In Italy it is found at Usseglio in Val di Lanzo (Piedmont) and Iglesiente in Sardinia.

Complex sulphides

From among the numerous examples of complex sulphides, many of which are tetrahedral, the most common and most interesting are listed below.

Proustite, Ag_3AsS_3, crystallizes in the trigonal system in rhombohedra or prisms, coloured scarlet or vermilion. The best crystals are found at Chanarcillo, Chile. Proustite is found in many of the silver districts of the western United States.

Pyrargyrite, Ag_3SbS_3, trigonal, is also dark red and because of this it is called, together with proustite, 'red

silver'. They are found in the same deposits and are often associated with one another.

Bournonite, $CuPbSbS_3$, and the other similar minerals: *boulangerite*, $Pb_5Sb_4S_{11}$; *jamesonite*, $Pb_4FeSb_6S_{14}$; *meneghinite*, $Pb_{13}Sb_7S_{23}$; are not easy to distinguish from each other at first sight. They are dark lead gray with a metallic lustre, a prismatic and often acicular habit and they usually form felted aggregates (especially boulangerite). They are frequently found in lead deposits, as at Trepča in Yugoslavia and Bottino near Saravezza in the Apuanian Alps (a famous place for meneghinite) in Italy.

Cannizzarite, $Pb_3Bi_5S_{11}$, is a rare mineral produced by volcanic gases; it forms small, pale gray, lustrous crystals in the lava of Vulcano (Aeolian Islands).

Sartorite, $PbAs_2S_4$, and *jordanite*, $Pb_{14}As_7S_{24}$, are two of the numerous sulphides which can be found in dolomite at Binnatal in Switzerland. They have a tabular or prismatic habit, lead-gray colour, and metallic lustre.

Realgar, orpiment

Realgar, AsS, is monoclinic; it forms compact masses or not very large crystals, red or orange, with a resinous lustre and low degree of hardness, (1.5–2). When exposed to air and light it changes to granular orpiment. Well-known localities for this mineral are Cavnic, Baia-Sprie, and Săcărimb in Romania, Binnatal in Switzerland, and Matra in Corsica.

In the United States, realgar is found at Mercur in Utah, and Manhattan, Nevada; it is deposited from the geyser waters in the Norris Geyser Basin, Yellowstone National Park.

Orpiment, As_2S_3, is also monoclinic; it only occasionally forms crystals, more often earthy, foliated or fibrous masses coloured lemon yellow, gold, brownish yellow, or orange yellow on a fresh cleavage surface. It is a common alteration product of realgar and tends to disintegrate when exposed to air and light. Orpiment was used as a pigment in ancient times, and large-scale deposits exist in Georgia (USSR), Turkey, and Persia.

Right: Realgar in dolomite with pyrite. Binnatal (Switzerland)

Above left: Bournonite on siderite. Huttenberg (Austria)

Left: Jordanite in dolomite. Binnatal (Switzerland)

Class III: halides

These are compounds of metals with the halogens (fluorine, chlorine, bromine, and iodine). Generally they have a low degree of hardness and low specific gravity; their lustre is vitreous, and the colour can vary greatly, as in the case of fluorite. Some halides have considerable importance in industry.

Halite, NaCl: isometric system; sg 2.16; hardness 2.

This mineral, which is essential to human and animal life, is usually found in compact or granular masses, or in cubic crystals, colourless or white, violet, yellow, blue, or purple. It shows a well-defined cubic cleavage and vitreous lustre. The crystal faces often show extra growth at the edges, which results in curved faces in a hopper-like formation. There are large deposits in several parts of the world: Stassfurt in Germany, Wieliczka in Poland, Salzburg in Austria, etc. Thick halite deposits occur in a number of American states, including New York and Michigan.

Sylvite, carnallite

Similar to rock-salt in colour and shape, though much more rare, is *sylvite*, KCl, which occurs in cubic and octahedral forms, or in compact masses, in the deposits of Stassfurt, Germany, and Carlsbad, New Mexico. It is often associated with *carnallite*, $KMgCl_3.6H_2O$, orthorhombic, which very rarely occurs in separate crystals, more commonly in milk-white or reddish granular masses.

Fluorite, CaF_2: isometric system; sg 3.18; hardness 4.

This has probably the widest range of colours of any mineral: from clear colourless to deep violet, green to yellow, sky blue to pink. There can be magnificent crystals in large cubes or octahedra, and there is a clear cleavage parallel to the octahedral face, present in both crystals and masses; the lustre is vitreous. Good crystals are found in various parts of Europe, such as Germany and England. In Italy, clear colourless cubes of fluorite are found at Corvara

Right: Cubic crystals of fluorite with quartz. Durham (England)

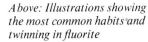

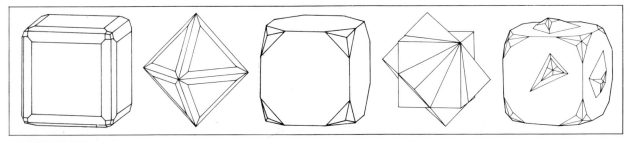

Above: Illustrations showing the most common habits and twinning in fluorite

Left: Halite. Santa Caterina Vallermosa (Sicily)

(Bolzano). In Switzerland it occurs in pink octahedra in the dolomite of the St Gotthard region. Some United States localities are Westmoreland, New Hampshire; Trumbull, Connecticut; and Macomb, New York, where very large sea-green cubes are found. Fluorite has been mined in the United States principally in western Kentucky, and Hardin and Pope counties, Illinois.

Cryolite, atacamite, cotunnite, nadorite

Cryolite, Na_3AlF_6, has a similar appearance to the white compact type of fluorite, and is monoclinic, only occasionally appearing in pseudo-cubic crystals. It is found almost exclusively in the Ivigtut deposits of Greenland.

An interesting chloride of copper is *atacamite*, $Cu_2Cl(OH)_3$, generally occurring in aggregates of slender prisms coloured dark green; it comes mainly from Chuquicamata in Atacama (Chile), from where it takes its name.

A rarer mineral is the orthorhombic *cotunnite*, $PbCl_2$, found in colourless crystals in lava from Mount Vesuvius (Italy); and equally rare is *nadorite*, $PbSbO_2Cl$, in brownish-yellow tabular or lenticular crystals occurring at Djebel Nador in Algeria.

Class IV: oxides and hydroxides

Compounds of the metallic elements with oxygen are placed in this group and they are extremely varied in their appearance and physical properties. Many have a secondary origin, that is, they are formed as a result of changes that have occurred to other minerals, such as limonite from iron-bearing minerals, stibnite from antimony-bearing minerals, etc.

It is a curious fact that *ice* (H_2O) belongs to this class: it crystallizes in the hexagonal system at a temperature of between 0°C and −80°C, and into the isometric system at a rather lower temperature.

Cuprite, Cu_2O: isometric system; sg 6.14; hardness 3.5–4.

It is an important secondary ore of copper and forms in the zone of weathering of copper deposits in association with malachite and azurite. It can be found in compact masses, elongated parallel growths known as 'plush copper' or *chalcotrichite*, or in octahedral crystals, dark red in colour and with a semi-metallic to opaque lustre. Fine crystals can be found in Redruth and Liskeard in Cornwall, England; Bisbee and Clifton in Arizona, and Del Norte County, California; Chessy near Lyons, France; Tsumeb, South West Africa. In South Australia it occurs in the Burra district, and it comes from Broken Hill in New South Wales. In Bolivia it is found at Corocoro, La Paz, and in Chile fine crystals come from Andacollo, south of Coquimbo.

Tenorite, zincite

The rare *tenorite*, CuO, is a triclinic copper oxide found in lava from Mount Vesuvius in thin layers of a very brilliant black. When found in compact

Below: Octahedral magnetite. Kollergraben, Binnatal (Switzerland)

masses, showing a conchoidal fracture and pitch-black lustre, or in granular aggregates, black in colour, it is called *melanochalcite*.

Zincite, on the other hand, is a hexagonal zinc oxide, of a yellow-orange or dark-red colour, transparent or translucent, with a sub-adamantine lustre. It is found in the famous Sterling and Franklin mines in New Jersey. It also comes from Olkusz, Poland; near Paterna, Almeria (Spain); and it is found in the Heazlewood Mine, Tasmania.

Spinel

This name denotes a group of minerals all of which have very similar characteristics. There are about a dozen different varieties, some of which are of considerable economic importance; crystallization is isometric, generally in octahedra.

The true *spinel*, $MgAl_2O_4$, is a fairly common mineral with a colour that ranges from red to blue to green to colourless. It has a glassy lustre, a specific gravity of 3.55, and a hardness of 7.5–8. The transparent red variety known as *ruby spinel* or *balas ruby* is used as a gem. The iron spinel, dark green to black, is called *pleonaste*, and the brownish-coloured chrome spinel is called *picotite*.

Crystals that can be used as gems come mainly from the stream sands of Ceylon, Burma, and Madagascar; large octahedral crystals from New Jersey.

Very similar are crystals of *hercynite*, $FeAl_2O_4$, and *galaxite*, $MnAl_2O_4$, which are black in colour, and *gahnite*, $ZnAl_2O_4$, dark green or dark blue, such as those found in Tiriolo, Calabria (Italy). Spinels of this type occur in large crystals in deposits at Sterling and Franklin, New Jersey, and at Rowe, Massachusetts.

Of considerable importance because it is so rich in iron, is *magnetite*, Fe_3O_4, so called because it is strongly magnetic (a fact well known to the Ancient Greeks). The octahedral or rhombohedral crystals are black with a metallic lustre, and in common with all other spinels, it has no cleavage at all. Its specific gravity of 5.17 is the highest of all members of this family, and its hardness is 6. Magnetite can contain chromium or manganese.

The largest deposits of this common and widespread mineral are found in Sweden (Kirunavaara, Gällivare, and Luossavaara), and in the Urals (Mount Magnitnaya). In the United States magnetite has been found in commercial quantities in several localities in the Adirondack region in New York, and in Utah, California, New Jersey, and Pennsylvania.

A zinc- and iron-rich variety of spinel is called *franklinite*, $(Zn,Mn,Fe'')(Fe''',Mn)_2O_4$, and its name derives from the fact that it is found in Franklin, New Jersey, where its black octahedral crystals embedded in calcite are associated with willemite, zincite, garnet, etc.

Trevorite, $NiFe_2O_4$, is a metallic black variety of spinel containing nickel, found in the Transvaal.

Multiple twinned chrysoberyl. Brazil

Chromite, $FeCr_2O_4$, is the only chromium ore to be used in industrial production. It is usually found in black granular masses and is always associated with serpentinite and peridotite rocks in Rhodesia, in the Urals, and in Turkey. Minute crystals are found in the serpentinites near New York (Hoboken and Staten Island), and in Maryland. Small, but economically workable, deposits have been found in Maryland, North Carolina, and California. During the Second World War, bands of chromite were mined in the Stillwater igneous complex, Montana.

Chrysoberyl

Chrysoberyl, $BeAl_2O_4$, is distinguished by its hardness (8.5). It can be present in pegmatite dikes, and occurs in orthorhombic crystals which are tabular and commonly twinned, with asparagus-green, grass-green, greenish-brown, or yellow colouring. The highly valued variety, *alexandrite*, which is used as a gem stone, takes on a violet colour in artificial light, while the *cat's eye* variety is notable for the light reflections it generates when cut to a curved shape.

Very high quality *alexandrite* is found in the area of the river Ai near Sverdlovsk in the Urals, with examples of somewhat lesser importance in Ceylon and Madagascar. Transparent or translucent twinned varieties of a greenish-yellow colour occur in Espirito Santo in Brazil. In the United States it is found in pegmatites in Oxford County, Maine, at Ragged Jack Mountain, Hartford, and at a number of other localities.

*Right: Corundum in
zoisite rock.
Mt Matabatu
(Tanzania)*

Corundum, Al_2O_3: trigonal system; sg 4–4.1; hardness 9.

This is a very important mineral for two main reasons: first, because it has been used since the earliest times as a precious stone in the red-coloured transparent variety (*ruby*) and the blue variety (*sapphire*), and secondly, because the compact grayish type called emery is used for industrial purposes as an abrasive and as a refractory material. The crystals, usually showing a hexagonal section and tapered at the extremities, show a cleavage across their length, and have an adamantine lustre. The colour varies from red to blue, yellow, green, and gray; under ultraviolet light it is frequently a fluorescent red or orange. The finest gem stones are found in Burma, Thailand, and Ceylon, and those of rather lesser value come from Australia and the United States. Red corundum in a green rock rich in zoisite can be found in Tanzania. Emery is mined at Náxos and Sámos in the Aegean, in Turkey, and in the United States, where common corundum is found in various localities along the eastern edge of the Appalachian mountains in North Carolina and Georgia. Small sapphires of fine colour are found in various parts of Montana.

*Below: Hematite. Rio
Marina, Elba (Italy)*

Hematite, Fe_2O_3: trigonal system; sg 5.26; hardness 5–6.

This oxide, the name of which derives from the

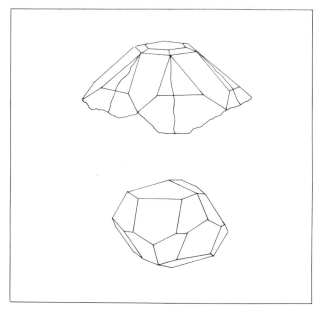

Greek word meaning 'blood' in allusion to the colour of the powdered mineral, is an important iron ore. It is found in earthy, ochrous masses, or else in concretionary (botryoidal), fibrous, micaceous, or scaly masses, and above all in aggregates of lamellar or lenticular crystals or even more complex forms. The colour is generally iron black with a metallic lustre, with brownish red as a variant. There are even some crystals covered in an iridescent patina.

Aggregates of thin plates (lamellar crystals) grouped in rosettes are known as 'iron roses' and are very common in conjunction with adularia, quartz, and chlorite. It is also called *oligist* which means 'very little' in Greek and is thus referred to because it has a low iron content as compared with magnetite. Extremely famous are the many-faced crystals found in Rio Marino on the island of Elba, Italy, and the foliated, rhombohedral crystals whose high lustre has caused them to be called specular, found in Brazil. In the United States, the Mesabi Range of Minnesota yields a fair quantity of rather small crystals and in Michigan can be found micaceous schist-like hematite with a brilliant lustre.

Ilmenite, perovskite, stibiconite, betafite

Ilmenite, $FeTiO_3$, trigonal, rarely occurs in true crystals but more often in granules and large scales, with a black powder. This mineral is somewhat similar to the preceding one because of its black, lustrous, and tabular aspect. It is a common accessory mineral in gabbro, diorite, and anorthosite rocks. Good crystals are found in Krågerö and Arendal in Norway, Binnatal in Switzerland, Bourg d'Oisans in France, and the Malenco Valley in Lombardy and Ossola in Piedmont, both in Italy. In the United States it can be found in the states of Washington and Connecticut, in Orange County, New York, and it is actively mined at Tahawus, Essex County, also in New York.

Below: Clear quartz. Sasso Nero, Bolanzo (Italy)

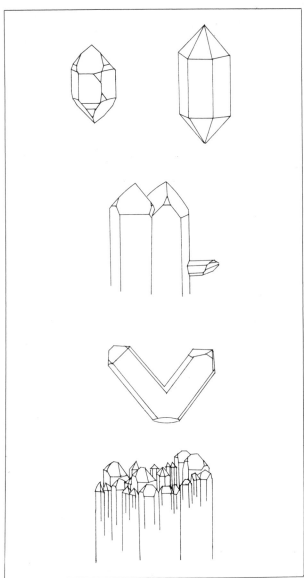

Prismatic forms, regular and twinning characteristics of quartz

An interesting and somewhat rare mineral is *perovskite*, $CaTiO_3$, which can be orthorhombic or isometric and has pseudo-cubic crystals of a black, brown, or yellow colour. Excellent black crystals are obtained in the Urals and brownish-yellow ones embedded in calcite are found in the Malenco Valley in Italy, while darker crystals are mined in the Vizze valley and at Piossasco, again in Italy.

A variety known as *stibiconite*, $SbSb_2(O,OH,H_2O)_7$ is of a yellowish-earthy colour due to its association with antimony, which often replaces it. Individual groups of antimony crystals now entirely changed to stibiconite are found in San Luis Potosí, Mexico, and Poggio Fuoco in Tuscany, Italy.

Betafite, isometric, is an oxide of calcium, uranium and titanium with rare earths, and has octahedral pitch-black crystals with a glassy lustre and conchoidal fracture, generally covered with a yellowish patina. It most celebrated place of origin is Madagascar.

Right: Cavity lined with crystals of amethystine quartz. Rio Grande do Sul (Brazil)

Silica group: quartz, tridymite, opal

This group comprises one of the most abundant and widely disseminated on the Earth's crust. Quartz, above all, is an essential constituent of volcanic, metamorphic, and sedimentary rock.

Quartz, SiO₂, crystallizes in the trigonal system and is found in the well-known prismatic pyramid configuration. Usually glassy or milky in colour (rock crystal), it can also show a brown coloration (smoky quartz), black (morion quartz), yellow (citrine), violet (amethyst), pink, blue, and green.

There are frequently other minerals that occur as inclusions in quartz, such as rutile (rutilated quartz), tourmaline, mica (aventurine quartz), blue alternating with yellow crocidolite (falcon's eye and tiger's eye). Quartz has a glassy lustre and conchoidal fracture, and is seventh in the Mohs hardness scale; the prism faces are finely striated across their length. Clear quartz is found in many parts of the world, the most celebrated examples being the Dauphiné groups in France, the big crystals with tourmaline in Brazil, the perfect clear crystals in the marble from Carrara in Italy, and the druses with magnesite in Dosso dei Cristalli in the Malenco Valley, Italy. The amethyst, citrine, and rose quartz, other common crystalline varieties, are usually mined in Brazil.

Chalcedony is a microcrystalline variety of quartz, found in concretionary forms, translucent, and of a white, black, or gray-blue colour. Chalcedony surfaces banded in several colours are clearly seen in the type

Above: Smoky quartz. Val Giuf, St Gotthard (Switzerland)

Right: Agate chalcedony. Brazil

Far right: Layered jasper. Siberia

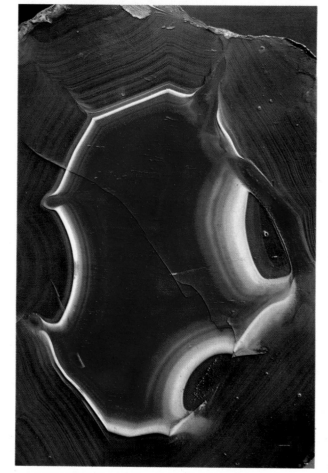

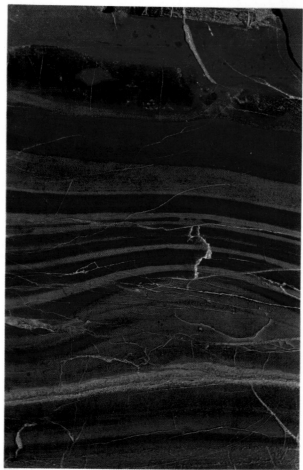

Left: Agate chalcedony with carnelian nucleus. Brazil

31

called *agate*, while in *onyx* the colours are strongly contrasted, usually in black and white, and the banding is straight. There are innumerable other names which have been assigned to the smallest variations in the colour and formation of the members of this group, but we do not have the space to deal with them all in this volume. A great many of these stones come from Brazil and Uruguay, and they are widely used as ornaments.

Jasper is yet another variety of quartz, and is a compact microcrystalline variety rich with impurities and therefore variously coloured green, red, yellow, brown, and black. As it is fairly common and has an attractive appearance, it was used as a gem many hundreds of years ago, and especially during the Renaissance. Brown and red jasper were obtained from Egypt, while other colours came from the Tuscan Apennine mountains in Italy.

Tridymite has the same composition as quartz, but it crystallizes in the orthorhombic and hexagonal systems. It is only found in volcanic rock, as for example in Eifel in Germany, Puy-de-Dôme in France, and Padua in Italy.

Opal, $SiO_2 \cdot nH_2O$, is a hydrous silica in a colloidal state, and therefore rather than forming crystals it exists in compact masses, veins or amygdales; it can

Precious opal.
Queensland (Australia)

Far left: Silicified wood. Egypt

Left: Fire opal. Zimapan (Mexico)

Needle-like rutile with chlorite. Valle Aurina, Bolzano (Italy)

Below: Twinned cassiterite (wood-tin). Horní Slavkov, Bohemia (Czechoslovakia)

be colourless, white, blue, gray, rose, or green. *Precious opal* is distinguished by its characteristic iridescence and is found widely in Australia. *Fire opal* has intense red reflections and comes from Zimapán in Mexico. A white, milky opal is found in association with magnesium in seams at Caselette and Baldissero Canavase, Piedmont, Italy. *Hyalite* is a white, transparent, glassy opal, with a green fluorescence under ultraviolet light and it is found in granite in several locations. *Wood opal* is formed from tree trunks that have been completely transformed into silica, and is found throughout the United States, with especially large quantities in Clover Creek, Idaho; it also exists in Egypt and Brazil, and examples of very large silicified tree trunks have been found in Lake Omodeo in Sardinia.

Rutile, TiO_2: tetragonal system; sg 4.23; hardness 6–6.5.

The crystals can be prismatic, compact, and longitudinally striated, or slender and needle-like, or latticed, or in filiform clusters. Colours are brown, black, red, yellowish, bluish, or violet, with metallic adamantine lustre. It frequently forms 'elbow twins', so-called because of the angle formed between the two intergrown crystals. A variety called *sagenite* is copper-red coloured and consists of foliated crystals which are

either single or interlaced to form a latticed surface, and are embedded in quartz.

Rutile is a widespread mineral in igneous and metamorphic rocks, and is commonly concentrated in auriferous sands. It is found throughout the European Alps, and in the United States large crystals are found at Graves' Mountain, Lincoln County, Georgia, and fine crystals come from several places in North Carolina. It also comes from a number of localities in Pennsylvania, Connecticut, New York, and Massachusetts.

Cassiterite, SnO_2: tetragonal system; sg 6.99; hardness 6–7.

It is virtually the only tin mineral to be used industrially. The Phoenicians obtained it in the legendary Cassiteride Islands, which were perhaps the Scilly Isles off the Cornish coast, and traded it throughout the Mediterranean region. It occurs in granular masses or prismatic crystals, of a yellowish, reddish-brown, or brown-black colour, with a greasy, adamantine lustre. The miner's term 'visor tin' describes the very common elbow-shaped twins with a characteristic notch. Large deposits are found in Bolivia and Indonesia, while the best examples of twinned crystals are found in Cornwall (England), Zinnwald (Saxony), Panasqueira (Portugal), and Katanga (Congo). Several other localities in Spain and Portugal provide fairly good specimens. In the United States cassiterite occurs in pegmatites (of no commercial value, although vein deposits in Virginia and California have been unsuccessfully worked).

Pyrolusite, psilomelane, wad, manganite

The oxides and hydroxides of manganese are very numerous, widely disseminated, and often so closely associated with each other that it is difficult to distinguish the different varieties at first sight, especially as they occur in earthy masses. Some of the most common are the following:

Pyrolusite, MnO_2, forms in compact masses which are fibrous or earthy, iron black or steel gray in colour, or else black dendritic or arborescent shapes on the fracture surface of various rocks. When its crystals are stubby and prismatic it is called *polianite*. It is an important manganese mineral, highly prized in the steel industry.

Psilomelane, $(Ba,H_2O)_2Mn_5O_{10}$, is found only in compact masses which are mammillary, stalactitic, opaque, earthy, and of a blackish colour. Psilomelane and pyrolusite are frequently associated with goethitite, limonite, braunite, etc.

The generic name of *wad* is given to a substance formed from several different hydrous manganese oxides, without any constant composition and therefore more of a rock than a mineral. It is found in earthy or concretionary masses of a black or brownish-black colour.

Manganite, $MnO(OH)$, crystallizes in the orthorhombic system in prismatic striated crystals of a dark steel-gray or iron-black colour. Fine crystals are found in Ilfeld in the Harz Mountains, Ilmenau in Thüringen, and Siegen in Westphalia (Germany).

Anatase, brookite

Anatase, previously known as *octahedrite*, has the same chemical composition as rutile and crystallizes into the same system, but it has a lower specific gravity (3.90), is yellowish, brown, or black, and has a decidedly pyramidal or bi-pyramidal habit. The lustre is metallic or glassy in the translucent variety. Like rutile it is found mainly in the metamorphic rocks in the Alps, in Dauphiné in France, in Binnatal and St Gotthard, Switzerland; in the United States it is found at Somerville, Massachusetts, and Magnet Cove, Arkansas.

Similar to the previous mineral in its chemical composition, but belonging to the orthorhombic system, is *brookite*, always appearing in lustrous crystals with a tabular or prismatic habit and a yellowish-brown, pinkish-brown, or iron-black colour. It is found in association with the previously mentioned minerals in crystalline schists in the Alps.

Wolframite, $(Mn,Fe)WO_4$: monoclinic system; sg 7.25; hardness 4.5.

Right: Wolframite with zinnwaldite. Panasqueira (Portugal)

Left: Anatase (or octahedrite) coloured brown. Kollergraben, Binnatal (Switzerland)

This is an important member of the tungstate group and has stubby crystals of a prismatic or tabular habit, or else it is found in bladed masses revealing a perfect cleavage and semi-metallic lustre; the colour is dark gray, brownish black, or iron black. If the crystalline form is absent it can be confused with cassiterite, though there is no cleavage in the latter.

Most examples originate in China, the United States, Portugal, England, and Bolivia. The highest quality crystals are obtained from the tin districts of Saxony and Bohemia, Panasqueira in Portugal, and Llalagua in Bolivia. Wolframite occurs in the United States mainly in the Black Hills, South Dakota.

Columbite, euxenite

Among the rarer oxides is *columbite* (Fe,Mn)Nb, Ta_2O_6, forming a series from the pure *niobate columbite* (oxide of iron, manganese, and niobium – the latter is known in the United States as *columbium*) and *tantalite* (oxide of iron, manganese, and tantalum); it presents prismatic or tabular crystals, coloured iron black, pitch black, and with a metallic lustre. Interesting examples come from the Scandinavian countries, the United States, Congo, and Australia. In the United States it is found at Standish, Maine; Haddam, Middletown, and Branchville, Connecticut; in Amelia County, Virginia; Mitchell County, North Carolina; Black Hills, South Dakota; and near Canon City, Colorado.

Euxenite contains uranium among its other elements. It is black, has a greasy or glassy lustre, and shows a conchoidal fracture. It is abundant in Switzerland, Norway, and above all in Madagascar from where fan-shaped groups of crystals are obtained.

Uraninite, UO_2: isometric system; sg 10.63; hardness 5–6.

This extremely important uranium mineral forms either cubic or octahedral crystals or a compact mass, with a glassy aspect and a banded structure. The colour is invariably black, the lustre pitchlike or greasy. It is strongly radioactive. The largest veins are found in the United States, Canada, and Katanga (Congo). Fine cubic crystals come from Wilberforce in Ontario (Canada), Shinkolobwe in Katanga, and Jáchymov in Bohemia (Czechoslovakia). In the United States it occurs at Middletown and Portland and other localities in Connecticut; near Central City, Colorado; and in Llano County, Texas.

Curite, gummite

Curite, $Pb_2U_5O_{17}.4H_2O$, is one of the numerous hydrous oxides and hydroxides of uranium. This reddish-orange mineral is found in compact earthy masses, and the main source is Katanga.

Gummite, however, is a mixture of various hydrous oxides of uranium, lead, and thorium. Like bauxite, limonite, and wad, it should not be classified as a mineral, but rather a rock (an aggregate of several

minerals). It is yellow, orange, reddish, brown, or black, and is frequently found in association with uraninite, as it is a late stage in the alteration of the latter.

Brucite, $Mg(OH)_2$: trigonal system; sg 2.39; hardness 2.5.

This mineral is frequently found associated with serpentinites, or metamorphosed magnesian limestones. It has flattened crystals which are rarely prismatic, with a perfect basic cleavage, and commonly veined, with a fibrous, foliated, micaceous structure. The most common colour is white, but gray, blue green, and greenish yellow occur; the lustre is pearly; the fibrous variety, with some fibres as long as 70 centi-

Surface-altered euxenite. Ambatofotskyely (Madagascar)

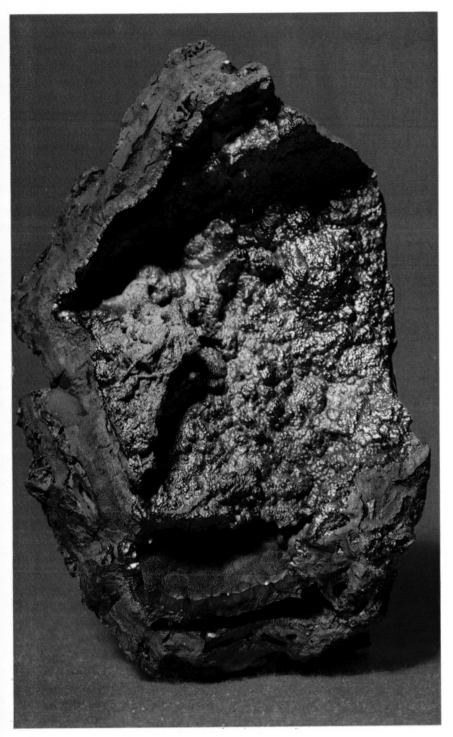

stalactite shapes are sometimes iridescent. As it is an alteration product of iron minerals, its geographic distribution is fairly wide, and excellent examples have been taken from the iron mines in Rio Marino, Elba, and also from Germany and Pennsylvania.

Bauxite

This is another mineral aggregate with an earthy aspect, and it is used as the raw material for the extraction of aluminium. Its colour varies from brick red to gray white, due to alterations in the sediment or volcanic rock in hot climates. It is a mixture of various aluminium oxides such as *boehmite, diaspore, hydrargillite, alumogel,* etc, mixed with iron hydroxides which impart the red colour, and with silicates. Extensive deposits exist in Yugoslavia (Istria, Dalmatia), France (Provence), South America (Guiana), and Italy, especially in Puglia. In the United States the chief deposits are found in Arkansas, Georgia, and Alabama.

metres (28 inches), is called *nemalite*.

Interesting examples have been found at Hoboken in New Jersey and at the Tilly Foster Mines near Brewster in New York State; it is also found at Crestmore, California, and in Lancaster County, Pennsylvania. If found in large quantities it is useful as an ore of magnesium.

Above: Concretionary goethite. Amberg (Pennsylvania)

Limonite

This name is generally given to a mixture of various hydrous oxides of iron, such as lepidocrocite, goethite, etc. Its aspect varies from stalactitic to mammillary, earthy, porous, or encrusted. When the material is concretionary, with a fibrous structure and a dark-brown or yellow-ochre colour, it is likely to be *goethite*. The

Right: Large brucite plates. Val d'Astico, Vicenza (Italy)

Class V: carbonates, borates, and nitrates

The most widespread minerals in this category belong to the *carbonate* group, those minerals with a chemical composition consisting of a metal combined with the anionic carbonate radical (CO_3). They usually have a medium or low hardness, they are sometimes white and sometimes highly coloured, and they can be transparent or translucent and usually form excellent crystals. The characteristic which differentiates them from the other categories is that they are soluble and effervesce in dilute hydrochloric acid, although for some carbonates the acid must be warmed first.

Borates are somewhat rare but can at times be found in quantities sufficient for industrial purposes. Most of them have a low specific gravity and a vitreous or greasy lustre; they are usually colourless, white, or gray, transparent or translucent; the hydrous compounds can easily become opaque and floury on contact with air, and frequently crumble away.

Nitrates will not be described in these pages because they rarely appear in Nature.

Magnesite

Magnesite, $MgCO_3$, is found in compact whitish masses or in rhombohedral and prismatic crystals which are either colourless, gray, or yellowish brown; it is translucent or opaque. Excellent lamellar or prismatic crystals of a yellow-brown colour, sometimes opalescent, cover hyaline quartz at Dosso dei Cristalli

 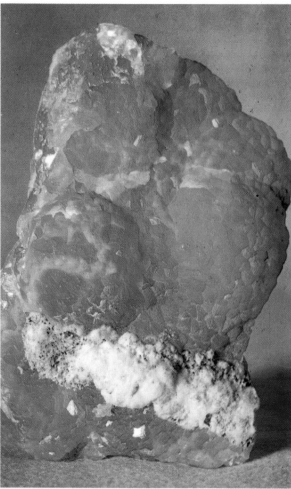

Far left: Magnesite on quartz. Val Malenco, Sondrio (Italy)

Left: Green mammillary smithsonite. Lavrion (Greece)

Right: Concretionary layered smithsonite. Montevecchio, Sardinia (Italy)

in the Malenco Valley, Lombardy (Italy).

Two ferrous varieties of magnesite are known as *breunnerite* and *mesitite*: the former occurs in rhombohedral, yellowish crystals in schist in Val di Vizze, Bolzano (Italy); and the latter is found in aggregation with lenticular yellow-brown crystals at Brosso and Traversella near Ivrea in Piedmont. In the United States the compact variety is found in serpentinite in the Coast Range, California. The sedimentary type is mined at Chewelah in Stevens County, Washington, and in the Paradise Range, Nye County, Nevada.

Smithsonite, $ZnCO_3$: trigonal system; sg 4.43; hardness 4-4.5.

This mineral, named after James Smithson, founder of the famous Smithsonian Institution, Washington, is widespread in zinc and lead deposits where it forms at the expense of blende. Normally appearing in a concretionary, stalactitic, or mammellar form, it is sometimes confused with hemimorphite (zinc silicate), and is known as 'calamine' by miners in Europe. It is also found in white or yellowish crystals, usually rhombohedral, translucent, and with a vitreous or semi-greasy

Above: Siderite in brown rhombohedra with the cleavage planes clearly visible. Neudorf, Harz (Germany)

Left: Brown siderite. Cornwall (England)

Right: Hexagonal siderite in which can be seen trigonal growth shadows, with quartz. Cornwall (England)

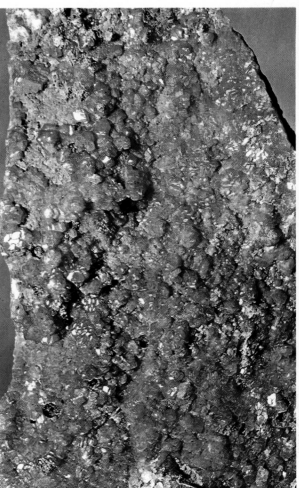

lustre. Concretionary forms, on the other hand, have a yellow, green, gray, blue, or milky-white coloration.

Fine examples have been obtained in Tsumeb (South West Africa), and Broken Hill (New South Wales), while mammellar, blue, concretionary masses exist in Lavrion (Greece), and Magdalena (New Mexico). A very beautiful, banded, bright-yellow smithsonite occurs at Masua near Iglesias, Sardinia (Italy), and green and yellow crystals are found at Monteponi, also in Sardinia. There is a notable occurrence at Nerchinsk, Siberia, and at Chessy, France. In the United States smithsonite occurs as an ore in the zinc deposits of Leadville, Colorado, and in Missouri, Arkansas, Wisconsin, and Virginia. Fine greenish-blue material has been found at Kelly, New Mexico.

A ferrous variety of smithsonite is known as *monheimite*, found at Montevecchio, Sardinia in small, yellowish, scalenohedral crystals.

Siderite, $FeCO_3$: trigonal system; sg 3.96; hardness 3.7–4.1.

Although this mineral is intensively mined in several places it is not as useful as magnetite and hematite. It is found in compact, spathic, granular masses, or in rhombohedral, lenticular crystals with good cleavability along the rhombohedral planes. It is yellowish brown, reddish brown, or even black due to oxidation of the manganese present. Fine brown crystals are obtained from Styria (Austria), Redruth and Camborne in Cornwall (England), and Morro Velho in Minas Gerais State (Brazil). There are extensive siderite veins in subAlpine Lombardy from Valsàssina to Valtrompia, and it is also found at Nurra and other parts of Sardinia. It occurs in many localities in the United States, such as the old mines at Roxbury, Connecticut; Vermont; New York State; in New England pegmatites; and in good crystals in Colorado ore veins.

Sphaerocobaltite, rhodochrosite

Sphaerocobaltite, $CoCO_3$, is rarely found in separate crystals but more frequently in rounded aggregates with a crystalline surface and radial structure, or otherwise in encrustations. The colour is rose to red with a tendency towards gray brown. The best examples are found at Libola in Liguria, Italy, and from Boléo, in Lower California.

Rhodochrosite, or *dialogite*, $MnCO_3$, is characterized by the various shades of rose colour which suffuse the crystals or compact concretionary masses. It is often used for ornamental purposes. When exposed to air it tends to turn brown black, as do all manganese compounds which are not already dark coloured. Groups of crystals are found in various parts of Colorado, such as Sweet Home Mine and John Reed Mine; also at Branchville, Connecticut, and Franklin, New Jersey,

Above left: Rhodochrosite (or dialogite). Cavnic (Romania)

Above: Sphaerocobaltite Musonoi (Katanga)

Above right: Prismatic calcite. Andreasberg, Harz (Germany)

Right: Scalenohedral calcite in parallel intergrowths. Korsnäs (Finland)

Above: Pink dolomite.
Seddas Moddizzis,
Sardinia (Italy)

Left: Ankerite.
Brosso, Piedmont (Italy)

as well as in the silver mines at Austin, Nevada.

Compact masses of an intense rose-pink colour and banded structure come from the province of Catamarca in Argentina.

Calcite, $CaCO_3$: trigonal system; sg 2.71; hardness 3.

This is one of the most common, most abundant, most easily found of all minerals. It can show various types of crystallization (700 basic varieties have been described), but the characteristic rhombohedral cleavage is always present. Its most common habits are rhombohedral, scalenohedral, and prismatic; it frequently forms stalactitic, concretionary, radial, or

crystalline calcite (chalk, limestone). The main districts for the occurrence of calcite in the United States are Joplin, Missouri, the Lake Superior copper mines, and Rossie, New York.

Dolomite, ankerite

Dolomite. $CaMg(CO_3)_2$: trigonal system; sg 2.85; hardness 3.5–4. This is the second most common carbonate on Earth (after calcite), mainly because it constitutes part of the large sedimentary rock formations which are mixtures of calcite and dolomite. It shows a constant rhombohedral form and its faces are often curved at both ends so acutely that they form

Pseudo-hexagonal aragonite crystals. Agrigento (Italy)

encrusted masses. Its lustre is vitreous, sometimes pearly and iridescent on the cleavage surfaces. It can be transparent, translucent, or opaque; colourless or yellow, brown, pink, blue, greenish, reddish, etc. It often becomes fluorescent pink under ultraviolet light. Concretionary calcite of a brown, red, or white colour is formed in rock cavities through the action of water (*stalactites, stalagmites*, some so-called 'onyxes'), or it forms incrustations such as *travertine*.

Good scalenohedral crystals, colourless or yellowish, clear or translucent, are obtained in Cumberland (England), and Mexico; in the Fontainebleau area (near Paris) aggregates of rhombohedral crystals are grayish white in colour and covered by sand; large hyaline crystals with a clear cleavage are known as 'Iceland spar' because of their occurrence in Helgustadir, Iceland.

The sequences of sedimentary rocks known as calcareous deposits are mainly composed of micro-

saddle-shaped crystals (*selliform aggregates*). It can be colourless and transparent, milky white or yellowish. There is a perfect cleavage parallel to the rhombohedral face, and it effervesces only in warm hydrochloric acid, whereas calcite does so in cold acid.

It occurs in excellent crystal forms in ore deposits such as those in Freiberg and Schneeberg, Saxony (Germany), Cornwall, and Keweenaw, Michigan, in association with the native copper. The pink or yellowish selliform crystals found with galena and calcopirite are well known at Joplin, Missouri, and at Picher, Oklahoma.

Ankerite, $CaFe(CO_3)_2$: trigonal system; sg 3; hardness 3.5–4. This is similar to dolomite, but brown. It is found in Austria and in Piedmont (Italy).

Aragonite, $CaCO_3$: orthorhombic system; sg 2.94; hardness 3.5–4.

Calcite and aragonite are both polymorphs of

calcium carbonate, the former trigonal, the latter orthorhombic. Because it is less stable, aragonite tends to invert to calcite under certain conditions. It forms in isolated acicular crystals, in radiating groups, in prismatic pseudo-hexagonal crystals consisting of an intergrowth of three individuals, in coral-like growths called *flos-ferri* (iron flowers), and in concretionary, fibrous encrustations. It can be colourless or gray, blue, green, reddish, or brown; sometimes transparent, it is more frequently translucent or opaque.

Extremely famous are the crystals with a prismatic, pseudo-hexagonal, multi-twinned formation and a gray or reddish hue, found in Molina de Aragón, Spain, the locality having given this mineral its name; equally famous are similar groups of a white, blue, and brown colour found in association with sulphur and calcite in the Sicilian sulphur mines and especially at Cianciana, Sicily. The *flos-ferri* are obtained from the siderite mines in Styria, Austria.

Strontianite, witherite

Strontianite, $SrCO_3$, has an orthorhombic structure and usually occurs in prismatic crystals, acicular or pyramidal, colourless, yellow, or brown. It is found in Germany (Drensteinfurt in Westphalia and Braunsdorf in Saxony). It forms granular masses at Schoharie, New York; associated with barite, pyrite, and calcite.

Witherite, $BaCO_3$, is also orthorhombic. It forms prismatic or dipyramid crystals which can be colourless, grayish, or yellowish, with a vitreous or greasy lustre on the fracture surface. The best specimens come from Fallowfield in Northumberland and Alson Moor in Cumberland, England; Rosiclare, Illinois; and S'Ortu Becciu near Dònori, Sardinia.

Cerussite, $PbCO_3$: orthorhombic system; sg 6.55; hardness 3.

This mineral, which is a lead ore, can form striated, reticulated crystals, or tabular twinned forms. It is rarely fibrous, often granular, and may be earthy or stalactitic. The colour is generally a pure or yellowish white, with an adamantine, greasy lustre. Good twinned crystals are found at Tsumeb (South West Africa); aggregates of reticulated crystals are mined in various parts of Iglesiente in Sardinia. In the United States they are found in Phoenixville, Pennsylvania; Leadville, Colorado; various districts in Arizona; in the Organ Mountains, New Mexico; and in the Coeur d'Alene district in Idaho.

Azurite, $Cu_3(CO_3)_2(OH)_2$: monoclinic system; sg 4.05; hardness 3.5–4.

This is a copper ore and forms in concretionary crusts or compact masses, with prismatic or tabular

Above: Cerussite, prismatic crystals. Tsumeb (South West Africa)

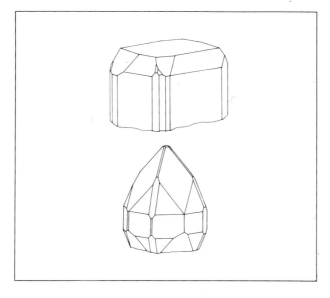

Phosgenite. Top: The most common type which ends in a pinacoid at the base. Below: A more rare form which is pyramidal in shape

crystals which are striated, very lustrous, and its colour ranges from intense blue to blackish blue. It also occurs in spherical aggregates. It can gradually become altered to malachite and one can find azurite almost entirely transformed (pseudomorphs of malachite after azurite). It is found in fine crystals at Tsumeb, South West Africa; Chessy, near Lyons, France; and Bisbee, Arizona.

Malachite, $Cu_2CO_3(OH)_2$: monoclinic system; sg 4.05; hardness 3.5–4.

It forms very easily, to the detriment of other copper ores such as chalcopyrite, bornite, cuprite, etc, and it

Below: Cerussite. Montevecchio, Sardinia (Italy)

is frequently found in association with azurite, chryso-colla, and brochantite. It forms in masses (sometimes of considerable dimensions) which can be concretionary, zoned, mammillary, botryoidal, or reniform. It can form radiating fibres or masses of needle-like crystals, and it can also have large prismatic crystals which are generally pseudomorphs after azurite. The lustre is vitreous in the needle-like crystal, glaze-like in concretionary masses, silky in some fibrous aggregates. The colour varies from pale to dark green, especially in the zoned variety which is used for ornamental purposes.

The most noted areas for compact malachite are Sverdlovsk in the Urals, Kilwezi in Katanga, Bwana Mkubwa in Zambia. Good pseudomorphs after azurite exist in Tsumeb. In the United States, it was formerly an important copper ore in the southwestern copper districts; at Bisbee, Morenci, and other localities in Arizona, and also in New Mexico.

Rosasite, aurichalcite

Rosasite, $(CuZn)_2CO_3(OH)_2$, is monoclinic and found in fine, fibrous crusts of a pale-green colour. It was first found at Rosas nel Sulcis in Sardinia, and later at Mapimí, Mexico.

Aurichalcite, $(ZnCu)_5(CO)_3(OH)_6$, is monoclinic and forms in tufts of delicate acicular crystals of a sky-blue colour and silky lustre. It occurs at Tsumeb; in Arizona and Utah; and at Yanga-Koubanza, Congo.

Hydrozincite, dawsonite, phosgenite

Hydrozincite, $Zn_5(CO_3)_2(OH)_6$, monoclinic, occurs in compact, undulating, powdery crusts of a white colour. It is found in zinc deposits and forms as a secondary mineral on calamine. Good examples are found in Masua, near Iglesias, in Sardinia. In the United States the right conditions are found in the south west and it is at its best in New Mexico.

Dawsonite, $NaAlCO_3(OH)_2$, orthorhombic, forms in white rosettes with a radiating fibrous structure, or in very flat, white, striated little crystals. It is found in Canada, Algeria, Tunisia, and Albania; also in several parts of Tuscany, Italy, especially around Mount Amiata.

Phosgenite, $Pb_2Cl_2CO_3$: tetragonal system; sg 6.13; hardness 2–3.

It is a rare secondary mineral of lead and is often found in association with cerussite. It usually has prismatic crystals which are often quite large, very rarely pyramidal, of a yellow-amber or grayish-yellow colour, with an adamantine or greasy lustre, transparent or translucent. It often shows a yellow fluorescence under ultraviolet light. Though it does occur in such varied places as Matlock in Derbyshire (England), Tarnowskie Góry (Poland), and Cluster County in Colorado, the most famous stones are mined at Monteponi, near Iglesias, Sardinia.

48

Bastnäsite, synchisite

Bastnäsite, $(LaCe)FCO_3$: hexagonal system; sg 4.9–5.2; hardness 4–4.5. Its colour varies from waxy yellow to reddish brown, with a greasy or vitreous lustre, opaque or translucent but rarely transparent. It forms in tabular crystals which are often quite large and arranged in a fan-like shape of individual crystals joined at one end. The name comes from the region of Bastnäs in Sweden where it was first found in 1838. It is especially common in Madagascar and the Congo. *Weibyeite*, which for example is found at Langesundfjord in Norway, is a pseudomorphosed bastnäsite of ancilite (a rare hydrous carbonate of strontium).

Synchisite is a fluocarbonate of calcium and cerium, and is orthorhombic or monoclinic, with small rhombohedral or tabular crystals, waxy yellow or grayish yellow, with a greasy or sub-adamantine lustre. It is found in syenite pegmatites in Greenland and also in the hydrothermal fissures in the Alps around Grigioni.

Northupite, pirssonite, gaylussite

The first two are rare sodium carbonates: *northupite*, a chlorocarbonate of sodium and magnesium, is isometric; *pirssonite*, a hydrous carbonate of sodium and calcium, is orthorhombic. They are found together in certain types of borate deposits such as the famous ones at Searles Lake in California. Northupite is colourless, in octahedral, gray or brown crystals;

49

Association of two tabular phosgenite crystals. Monteponi, Sardinia (Italy)

pirssonite has minute crystals which are white or colourless, a vitreous lustre and mostly prismatic crystals.

Gaylussite is a monoclinic carbonate of sodium and calcium with a higher water content than pirssonite. It is more common than the others, and forms long, fragile, colourless crystals which are transparent but tend to become opaque when dehydrated. It forms in closed lacustrine basins rich in sodium, where the climate is hot and dry, and it is frequently associated with soda, thermonatrite, pirssonite, and borax in, for example, Mongolia, Nevada, and California.

Thermonatrite, soda

Thermonatrite and *soda* (also called natron) are both hydrous carbonates of sodium, the second containing more water than the first, and they are typical of very alkaline lacustrine basins in hot climates. They occur in granular, efflorescent or encrusted crystals which are either opaque white or coloured gray due to clay impurities, and they are sometimes difficult to identify. They occur in many places, including California.

Nesquehonite

A hydrous carbonate of magnesium, it generally

forms crusts with a radiating structure, or it can assume forms which are concretionary, botryoidal, or tufts of white opaque crystals, while individual crystals are transparent. Though they were originally found in the coal-mines of Pennsylvania and France, these crystals are more easily formed by alterations in serpentinite rocks, such as at Kraubath in Styria (Austria), and various localities in Italy.

Hydrotalcite, stichtite, pyroaurite

Hydrotalcite, $Mg_6Al_2(OH)_{16}CO_34H_2O$, forms in flexible blade-like aggregates and white silky fibres which are slightly powdery, with pearly reflections. Its chemical composition explains why it is always found in association with rocks which are rich in magnesium, such as serpentinites. At Snarum in Norway, it is found in apple-green coloured serpentinites. It is also found at various localities in St Lawrence and Jefferson Counties, New York.

Stichtite, $Mg_6Cr_2(OH)_{16}CO_3.4H_2O$, is a rare mineral originating from changes in serpentinite formations, and is found in flexible scales or in lilac- or pink-coloured compact masses, in Tasmania, Morocco, and the Transvaal.

Pyroaurite, $Mg_6Fe_2(OH)_{16}CO_3.4H_2O$, forms isolated micaceous scales in a hexagonal system, or otherwise continuous layers of a lamellar or scaly form, coloured yellowish or brownish. It occurs in hydrothermal veins

Lilac-pink masses of compact stichtite. Bou Azzer (Morocco)

Right: Delicate tufts of acicular artinite. Val Malenco, Sondrio (Italy)

Far right: Fibrous-radiating ludwigite. Brosso, Piedmont (Italy)

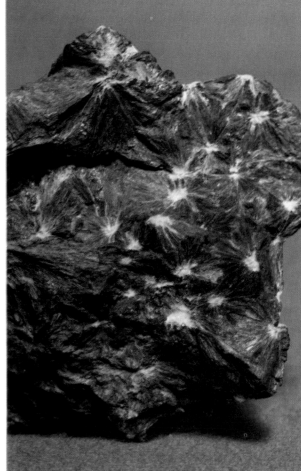

52

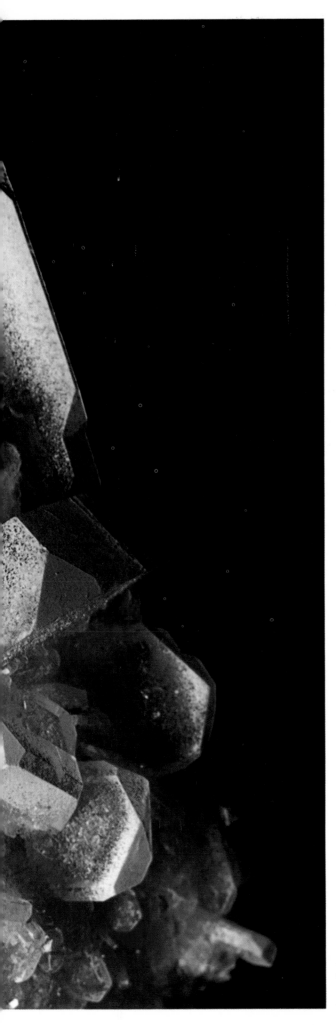

Tabular crystals of borax with a slight alteration to tincalconite. Searles Lake (California)

as at Langban in Sweden, or as a transformation of brucite in the serpentinites of Styria, Val Malenco, and Liguria in Italy, and in thin seams in serpentinite north of Fetlar, Shetland Islands, Scotland.

Rutherfordine, liebigite, schröckingerite

Rutherfordine, UO_2CO_3, is a rare radioactive mineral discovered in Tanganyika (now Tanzania) in 1906 and occurs in orthorhombic crystals which can be long, flat, or striated; or otherwise in rosettes and granular masses of microscopic needles; the colour ranges from yellowish white to yellow amber or yellow brown. It is not fluorescent under ultraviolet light. As well as in Tanzania, it is also found at Shinkolobwe in Katanga, and in Maine and New Hampshire.

Liebigite, $Ca_2U(CO_3)_4.10H_2O$, is another rare secondary mineral of uranium, whose existence has been known since 1848, when it was reported at Jáchymov in Bohemia; at that time it was called uranothallite. It is orthorhombic and forms stumpy prismatic crystals or fine crusts, scales, or even botryoidal masses. The colour is yellowish green and the mineral becomes fluorescent green under long or short wave ultraviolet light. It is a secondary mineral after pitchblende and is found in Saxony and Thüringen (Germany), Cornwall, and western Turkey.

Schröckingerite, $NaCa_3UO_2FSO_4(CO_3)_3.10H_2O$, can also be found with liebigite around pitchblende, of which it is a secondary ore; this too, was first reported in the famous Jáchymov mines in Bohemia. It forms globular aggregates or small, flattened, hexagonal crystals. The colour is greenish yellow of a more or less brilliant intensity; like autunite, it presents a vivid green fluoresence in ultraviolet light.

Hydromagnesite, artinite, brugnatellite

Two carbonates which are frequently in association with each other in serpentinite rocks are *hydromagnesite* $Mg_4(CO_3)_2.3H_2O$, found in small globular aggregates or white radiating rosettes with vitreous lustre; and *artinite*, $Mg_4(CO_3)_2(OH)_2.3H_2O$, in radiating tufts of delicate white acicular crystals with a silky lustre. They are found at Kraubath in Styria, Austria; in the United States, and in serpentinites in Italy. Both these minerals are alteration products of brucite in thermometamorphosed marble.

Sometimes found in association with these two minerals in the Malenco and Aosta valleys in Italy, is *brugnatellite*, $Mg_6Fe(OH)_{13}CO_3.4H_2O$, in lamellar powdery aggregates of a pinky-gray colour.

Ludwigite, vonsenite, ascharite

Ludwigite is a borate of magnesium and iron and forms fibrous aggregates, coloured dark grey or blue black. It is common in Hungary, Sweden, the United States, and Korea. In the United States it is found at Philipsburg, Montana, and in Utah.

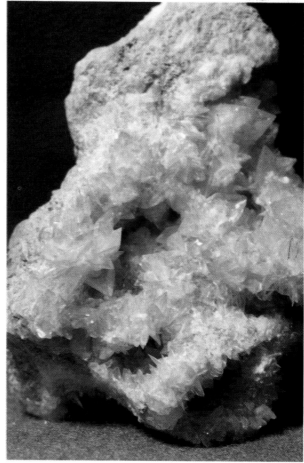

Vonsenite is a borate similar to ludwigite but richer in bivalent iron. It can be in fibrous masses, fibrous radiating masses, or needle-shaped crystals; the colour is greenish black or brown black. In Italy the variety known as *breislakite* in the volcanic rocks of Pozzuoli, Vesuvius, Vulcano, and Cimini is really vonsenite. It occurs at Riverside, California.

Ascharite and *szaibelyite* are hydrous borates of magnesium, which form in compact masses of an earthy white colour, together with ludwigite, in deposits in Saxony, and near Pioche, Lincoln County, Nevada.

Fluoborite

Fluoborite, $Mg_3BO_3(F,OH)_3$, presents hexagonal prismatic aggregates of small crystals arranged in tufts or in radiating structures, white or colourless. It occurs in contact with ludwigite and magnesite as at Norberg, Sweden, or with willemite, fluorite, and franklinite at Franklin, New Jersey.

Kurnakovite, inyoite, borax

Kurnakovite, a basic hydrous borate of magnesium, is found in compact or granular masses, white in colour, at Boron, California, and in Kazakhstan, USSR.

Inyoite has taken its name from Inyo County, California, where it occurs among several other borates. It is a hydrous compound of calcium, forming in compact masses or in prismatic and tabular crystals, transparent, colourless, or white due to dehydration.

Borax is a hydrous borate of sodium and occurs in compact masses or separate prismatic, colourless, translucent crystals. It is mined at Boron, California; in Tibet, and in Kashmir.

Loss of water can frequently cause almost complete alteration so that it becomes opaque, earthy, and transforms itself into tin-calconite, a borate with the same composition as borax but with fewer water molecules.

Inderite, meyerhofferite

Inderite, $Mg_2B_6O_{11}.15H_2O$, is monoclinic; it forms prismatic non-terminated crystals, rectangular in section and so deeply striated as to appear faceted, grayish white or colourless, translucent. It is found at Boron, California, and Lake Inder in Kazakhstan, USSR, after which it is named.

Meyerhofferite, $Ca_2B_6O_{11}.7H_2O$, is triclinic; it forms prismatic, flattened crystals which are sometimes rounded, vitreous, or fibrous silky masses, always white or colourless, and translucent. It is usually found in the borax formations of California and Turkey.

Ulexite, colemanite, kernite

Ulexite is a hydrous borate of sodium and calcium which forms in fibrous masses, frequently colourless and transparent. A curious optical phenomenon can be observed in transparent examples when a piece is taken consisting of two parallel faces cut transversely to the direction of the fibres and then polished: placed over any piece of writing, the image is completely transferred to the upper face and gives the impression that it is floating on a screen; it is for this reason that ulexite is also commonly known in the United States as 'television stone'. It is most commonly found in Boron, California.

Colemanite is a hydrous borate of calcium which often occurs in flattened, colourless, white, yellowish crystals, which can sometimes be coloured by impurities, with a vitreous lustre. It is found in association with other borates at Boron and Death Valley in California, Biğadiç (Turkey), and in Kazakhstan in the USSR.

Kernite is named after Kern County in California, where it was first discovered in the famous Boron deposits in 1926. It is chemically similar to borax, although it contains less water, and is not so soluble in cold water.

Hambergite, rhodizite, boracite

Hambergite, Be_2BO_3OH, is a rare orthorhombic borate typical of syenitic and alkaline pegmatites. It forms prismatic, flattened striated crystals which can be several centimetres long, with a good cleavage; its hardness is 7.5 and its specific gravity is 2.35; it is colourless, grayish or yellowish, with a vitreous lustre and good transparency. It is found in Norway, Madagascar, India, and California.

Rhodizite is a rare and complex isometric borate containing caesium, potassium, rubidium, aluminium, and beryllium. It forms rhombododecahedral crystals of 2–3 centimetres (roughly an inch) in diameter, in the pegmatites of Madagascar and certain parts of the Urals.

Lastly, mention should be made of *boracite*, a borate of magnesium containing chlorine, usually in isolated crystals with a cubic habit, colourless, white, yellow, and green; it forms in deposits of halite, gypsum, anhydrite, and potassium salts, such as the ones at Stassfurt in Saxony, Lüneburg near Hanover (Germany), and at Choctaw in Louisiana. It is also occasionally found in other parts of the United States of America.

Class VI: sulphates and related minerals

This class contains all minerals with the sulphate anion (SO_4); in addition, the chromates, molybdates, and tungstates are also included, though there are not more than twenty species in all. Tellurates, compounds of tellurium, also form part of this group, but are considerably less common, and for this reason are not described in this chapter. In general, sulphates are translucent or transparent, with a vitreous lustre and low degree of hardness. The most common varieties frequently have very beautiful crystals.

Thenardite, anhydrite

Thenardite, Na_2SO_4, orthorhombic, is found in an efflorescent form or in flat bipyramidal crystals, often twinned in a cross formation, grayish white in colour. It is obtained in Spain, California, Siberia, and Chile, and sometimes occurs in lava from Mount Vesuvius, Italy.

Anhydrite, $CaSO_4$, is rarely found in good crystal shapes, but more frequently occurs in compact saccharoidal masses of considerable dimensions. It is orthorhombic, forming equidimensional or tabular crystals, with a specific gravity of 2.98 and hardness of 3.5. When hydrated, it changes to gypsum while increasing its volume 60 times. It can be colourless, gray, blue, violet, or reddish. Crystals have been found at Bex in Vaud (Switzerland), and at the Simplon Tunnel and Stassfurt in Germany. *Vulpinite* is a gray compact anhydrite and is obtained from deposits of Vulpini in the lower Camonica Valley, Bergamo (North Italy).

Celestine, $SrSO_4$: orthorhombic system; sg 3.97; hardness 3–3.5.

Its name derives from the sky-blue colour it sometimes shows; it is found either in sedimentary rocks with sulphur, calcite, etc, or in the cavities of volcanic rock. The crystals are usually prismatic or tabular,

colourless, milky white, blue, or very occasionally reddish; they are sometimes transparent with a vitreous lustre, pearly on the cleavage surfaces. Splendid transparent crystals as much as 10 centimetres (4 inches) long have been obtained in the Sicilian sulphur mines, especially at Floristella (Enna). The rarer blue celestine comes from the same country. Pale-blue examples have been found at Matehuala, San Luis Potosí (Mexico), and at Portage, Wood County, Ohio. Of a quite disproportionate size – some are 40 centimetres (16 inches) long – are the blue tabular crystals which were found some time ago at Put-in-Bay, on Lake Erie, Ohio.

Barite, $BaSO_4$: orthorhombic system; sg 4.50; hardness 3–3.5.

It forms compact spathic masses or crystals which can be tabular, lenticular, prismatic, transparent, translucent, or opaque, of a white, yellow, brown, reddish, greenish, or blue colour. It has cleavages similar to those of calcite, but it can easily be distinguished from calcite due to its higher specific gravity and complete lack of effervescence when submerged in hydrochloric acid. It is the most common barium mineral and is used in the chemical, petroleum, and paper industries, among others.

Extremely beautiful crystals are obtained at Frizington and Dufton in Cumberland, England; at Freiberg and Marienberg in Saxony, Germany; and at Mies, Příbram, and Banská Štiavnica in Czechoslovakia. Red coloured 'rose' shapes are found at Norman in Oklahoma. It is also found at Cheshire, Connecticut; De Kalb, New York; and Fort Wallace, New Mexico. It occurs in large veins on Isle Royale, Michigan; barite enclosing grains of sand is found at Norman, Cleveland County, Oklahoma, and distorted crystals occupy cavities in fossil bones from the Bad Lands of South Dakota. Massive barite has been quarried in the United States in Georgia, Tennessee, Missouri, and Arkansas. At El Portal, California, at the entrance to Yosemite Park, barite is found in a vein with witherite.

Not nearly so common as barite, though its name is misleadingly similar, is *barytocalcite* which is not, however, a sulphate, but a carbonate of barium and calcium, and is monoclinic; it is found in some parts of England (Cumberland), Germany (Baden, Saxony), and Czechoslovakia.

Barytocelestine, on the other hand, is a mixture of barite and celestine and is to all intents and purposes a barite with a high strontium content.

Anglesite, $PbSO_4$: orthorhombic system; sg 6.38; hardness 2.5–3.

Originally found in Anglesey, Wales, it is a typical secondary mineral of lead deposits. It forms prismatic or tabular crystals which can be colourless and clear, or white, gray, blackish, yellow, and sometimes green or violet, with an adamantine or resinous lustre and conchoidal fracture. It can produce a yellow fluorescence under ultraviolet light.

Excellent crystals are found at Tsumeb (South West Africa), Sidi-Amor-ben-Salem in Tunisia, and Matlock in Derbyshire (England). In the United States large crystals are found in the Wheatley Mine, Phoenixville, Pennsylvania, and less well crystallized examples in the Missouri lead mines. Good crystals come from Coeur d'Alene, Idaho; the Castle Dome district, Arizona; and from Eureka in the Tintic district, Utah.

Jarosite, beudantite

Jarosite, $KFe_3(SO_4)_2(OH)_6$, is often ignored though it is fairly common; it forms powdery or earthy yellow-ochre coatings, or minute shiny crystals. When in its earthy form it is very similar to limonite, for which it is often mistaken. In formations containing pyrite it forms as a secondary mineral. It is found in many localities in Europe and North America.

When the potassium is replaced by sodium we have

Above: Prismatic-tabular anglesite. Iglesias, Sardinia (Italy)

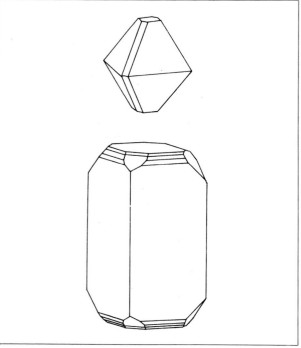

Anglesite. Above: A rare bipyramidal form from Nevada. Below: The more usual prismatic form

natrojarosite, replaced by silver it becomes *argentojarosite*, and replaced by lead *plumbojarosite*.

Beudantite is an arsenate with sulphate of lead and iron, which forms encrustations of minute crystals which can be rhombohedral or even pseudo-cubic; vitreous, black, brown, or dark green. It occurs in oxidation zones of lead deposits, as at Lavrion, Greece.

Brochantite, linarite, leadhillite, lanarkite

Brochantite, $Cu_4SO_4(OH)_6$, is orthorhombic; it forms dense felted interlacings of acicular crystals, emerald green or dark green in colour, with a vitreous lustre; it can easily be mistaken for malachite, but it does not effervesce in a cold hydrochloric acid solution. It is a secondary mineral common in copper deposits, as for example the one at Chuquicamata in Chile. Particularly well-crystallized specimens have come from Nevada, but it is also found in Arizona, New Mexico, Utah, and California.

Linarite, $(Pb,Cu)_2SO_4(OH)_2$, is also monoclinic; it occurs in aggregates of long, acicular, interlaced crystals of an intense blue, and is frequently in association with other secondary minerals of lead, copper, and zinc. It is abundant in Cumberland, England, and Tsumeb, South West Africa. Large crystals are rare, but some of the best have been found at the Mammoth Mine near Schultz, Pinal County, Arizona. Elsewhere it is usually in small crystals, as at Tintic, Utah; Eureka, Nevada; and Cerro Gordo, Inyo County, California.

Leadhillite, $Pb_4SO_4(CO_3)_2OH_2$, is monoclinic; it occurs as a secondary mineral in lead deposits, has pseudo-hexagonal crystals with a tabular habit, a vitreous or greasy lustre, (or sometimes pearly), and it is translucent. It was first found at Leadhills in Scotland, but it is also found at Granby, Missouri,

certain locations in the Tintic district, Utah, and in the Mammoth Mine, Arizona.

Sometimes, as at Leadhills, it is found in association with *lanarkite*, Pb_2SO_5, which is either compact or in long, gray or greenish-white crystals.

Alunite, hanksite

Very different from the preceding minerals is *alunite*, $KAl_3(SO_4)_2(OH)_6$, which forms extensive deposits in Utah and Colorado, and also central Italy (La Tolfa). It is found in compact masses, is granular in texture, and coloured white, yellowish gray, and reddish.

Hanksite, $KNa_{22}Cl(CO_3)_2(SO_4)_9$, has hexagonal, tubular or prismatic crystals which are usually colourless or yellowish gray, and is found in layers of salt at Searles Lake in California.

Hydrous sulphates

Some of these are known as 'vitriols': *chalcanthite* or copper vitriol, $CuSO_4.5H_2O$, forms sky-blue crusts; *melanterite* or iron vitriol, $FeSO_4.7H_2O$, is found in stalactitic masses of a green or blue-green colour; *goslarite* or zinc vitriol, $ZnSO_4.7H_2O$, can occur in efflorescences or fibrous stalactites of a whitish colour. All these are secondary minerals, that is, they form at the expense of others by incorporating molecules of water; they are common in ore deposits.

Below: Crust of small brochantite crystals. Sulcis, Sardinia (Italy)

Linarite, with crystals thickly interwoven. Sulcis, Sardinia (Italy)

The evaporation of water rich in magnesium causes the formation of *epsomite*, $MgSO_4 . 7H_2O$, also known as 'Epsom salts', which forms in white, very delicate tufts composed of acicular individuals; the floors of limestone caves in Tennessee, Kentucky, and Indiana are often covered with earth which is filled with minute crystals of epsomite.

A very common mineral is *morenosite*, $NiSO_4 . 7H_2O$, or nickel vitriol, which forms powdery coatings of a pale-green colour. It occurs, for example, in the serpentinites of Val Malenco (Lombardy) Italy.

Pickeringite, $MgAl_2(SO_4)_4 . 22H_2O$, is another secondary mineral of ore deposits, found in aggregates of very fine, silky, acicular crystals, or in encrustations, coloured yellowish white. It occurs in Thüringen,

Germany; the Salzburg district, Austria; and South America.

Halotrichite has a similar aspect and composition to the above, but it contains bivalent iron instead of magnesium. It is found in several mining areas in Italy, such as the sulphur deposits of Pozzuoli.

Coquimbite, $Fe_2(SO_4)_3 . 9H_2O$, forms prismatic or bipyramidal crystals with a hexagonal section, or granular aggregates, amethyst violet in colour, with a vitreous lustre. It is found in northern Chile, associated with several other sulphates, and in the United States it occurs at the Redington mercury mine, Napa County, California.

Alunogen, $Al_2(SO_4)_3 . 16H_2O$, triclinic, rarely forms prismatic crystals, but is mostly found in fibrous

60

Acicular interwoven gypsum crystals. Sicily (Italy)

masses, encrustations, or efflorescences, of a whitish colour. It is mainly formed by the alteration of iron pyrites in coal seams or rock fissures, and in alum shales. It has an acid, astringent taste.

Alunogen is found at Doughty Springs and Alum Gulch, Colorado, and extensive deposits occur in the Alum Mountains, Grant County, New Mexico.

Copiapite is a basic sulphate of bivalent iron, magnesium, and trivalent iron with 20 molecules of water, and takes its name from Copiapó in Chile. It forms orthorhombic, tabular crystals, or granular or scaly encrustations; the colour is sulphur yellow, yellow orange, or yellow green. It is found in the presence of many other sulphates, especially in pyrite oxidation zones, and therefore is often found in association with pyrite. In the United States it is found in copper mines such as the one at Jerome, Arizona; it has also been noted at Sulphur Bank, California, and in some of the California mercury mines, like Mount Diablo.

The alums are an important group of hydrous isometric sulphates containing sodium and aluminium (*sodium alum* or *natro-alum*), potassium and aluminium (*potash-alum*), ammonium and aluminium (*ammonia-alum*). All can be manufactured artificially in large and perfect octahedral or cubic crystals. When formed by Nature, however, they usually occur in white efflorescences in volcanic areas, clayey rocks which are usually rich in aluminium, and coal seams. They have a sweetish taste.

61

Astrakanite, also called blödite, is a hydrous sulphate of sodium and magnesium. It is monoclinic, and is found in stumpy little crystals, or granular or compact masses; it is colourless, bluish green or reddish due to impurities, and has a bitter taste. It is usually found in association with rock-salt, kainite, carnallite, polyhalite, anhydrite, etc. It occurs in several localities, in Germany, Austria, Poland, Chile, and the United States.

Polyhalite is a hydrous sulphate of potassium, calcium, and magnesium; triclinic, it is generally found in compact or fibrous masses, colourless or coloured by various impurities. It is a salt which frequently accompanies rock salt, anhydrite, etc, in deposits which have formed as a result of evaporation in marine basins, and as such it is present in almost all the salt deposits of the world.

Kainite is a hydrous chlorosulphate of potassium and magnesium; it is monoclinic and is also typical of salt deposits where it is found with rock-salt, carnallite, sylvite, etc. Rarely in crystals, it usually forms compact masses interstratified with other salts from which it is difficult to distinguish the kainite at first sight. The most notable localities are northern Germany, Poland, and the lower Volga (USSR).

Voltaite is a hydrous sulphate of potassium and iron found in fragile earthy crusts of a blackish green or olive green. It forms in volcanic areas (Pozzuoli,

Prismatic crocoite. Dundas, Tasmania (Australia)

Vesuvius, Vulcano in Italy), and as an alteration of sulphates (Cyprus, California, Chile).

Gypsum, $CaSO_4.2H_2O$: monoclinic system; sg 2.31; hardness 2.

Gypsum is a very common mineral and is easily identifiable because it is light and can be scratched by a finger-nail; in addition, it breaks naturally into thin transparent laminae, and when exposed to a flame it becomes an opaque white due to loss of water. The crystals are generally tabular or prismatic and when twinned they assume a 'swallowtail' or 'spearhead' appearance. It also occurs in fibrous or saccharoidal masses. Although it is usually colourless or transparent, it can also be gray, yellowish, or red; the lustre is semi-vitreous, pearly on cleavage surfaces. Imitative forms made up of lenticular crystals in rosette-shaped aggregates known as 'desert roses' are generally encrusted with grayish, yellow-brown, or reddish sand. *Selenite* is gypsum in flat, clear crystals; *sericolite* is the fibrous variety with a silky lustre; *alabaster* is compact, layered, can have various colours, and is used for ornamental purposes.

Gypsum deposits are extensively scattered throughout the world. In Europe, localities producing fine crystals are found in Greece, Czechoslovakia, Austria, Germany, Italy, Switzerland, France, Spain, and England. In the United States there are numerous localities for fine crystals including Lockport, New York; Mammoth Cave, Kentucky; Grand Rapids, Michigan; the Bad Lands, South Dakota. Commercial deposits are found in the states of Iowa, New Mexico, Kansas, Michigan, and New York.

Uranopilite, zippeite, johannite

These are the only three known sulphates of uranium. All contain molecules of water, but the only one to have other anions besides uranium is johannite, which contains copper.

Uranopilite forms crusts or globular aggregates of microscopic silky needles, coloured brilliant yellow or lemon yellow, showing a vivid green fluorescence under ultraviolet light. It is found in Bohemia, France, Cornwall, and Katanga.

Zippeite is also in the form of crusts of needle-shaped little crystals or earthy patinas; the colour is orange yellow, and fluorescence under ultraviolet light has green tinges. It is found mainly in Germany, Czechoslovakia, Cornwall, and Colorado.

Johannite shows prismatic crystals as well as patinas and fibrous aggregates; the colour is emerald green or apple green; there is no fluorescence.

Pyramidal scheelite crystal. Traversella Mine, near Ivrea, Turin (Italy)

Chromates, molybdates, and tungstates

Crocoite, $PbCrO_4$: monoclinic system; sg 6; hardness 2.5–3.

It occurs in prismatic crystals striated longitudinally, coloured hyacinth pink or orange, with an adamantine or vitreous lustre. No doubt the finest examples are found at Dundas in Tasmania, but it is also obtained at Sverdlovsk in the Urals and Conghonhas do Campo in the state of Minas Gerais, Brazil.

Vauquelinite, fornacite

These minerals are typical of oxidation zones of lead deposits containing chromium and arsenic. *Vauquelinite* is a phospho-chromate of lead and copper and forms minute monoclinic wedge-shaped crystals or mammillary fibrous aggregates whose colour varies from apple green and olive green to very dark brown. It is found in the Urals, Scotland, Brazil, Tasmania, and the United States.

Fornacite is a rare basic chrome-arsenate of lead and copper which forms crusts of minute vitreous olive green crystals, usually with dioptase, in the Congo.

Powellite, ferrimolybdite

These two minerals are usually formed as an alteration of molybdenite. *Powellite*, $CaMoO_4$, has tetragonal crystals, but it is generally found in earthy patinas or straw-yellow, greenish-yellow, or pale-yellow coatings.

It becomes fluorescent cream yellow or gold in ultraviolet light. Deposits are numerous, occurring in the Urals, Morocco, and the United States.

Ferrimolybdite, $Fe_2(MoO_4)_3.7H_2O$, occurs in fibrous crusts or tufts of canary-yellow or greenish-yellow tiny needles. It is widespread in all molybdenite deposits, for example in Sardinia and some parts of Calabria (Southern Italy).

Scheelite, $CaWO_4$: tetragonal system; sg 6.10; hardness 4.5–3.

It is an important tungstate mineral, usually fluorescent in white blue under ultraviolet light. It occurs in compact masses or bipyramidal crystals coloured honey yellow, yellow brown, greenish, or reddish gray. Transparent, translucent, or opaque, it has a vitreous lustre. Good crystals have been found in Korea, Malaysia, and Spain. In the United States, scheelite is mined near Mill City and Mina, Nevada; near Atolia, San Bernardino County, California; and lesser amounts are obtained in Arizona, Utah, and Colorado.

Stolzite

It is a tetragonal lead tungstate which forms in oxidation areas of some lead deposits with limonite, vanadinite, wulfenite, etc. The crystals have a

bipyramidal habit, are coloured red to brownish red to straw yellow, with a resinous lustre. It is found in Germany, Austria, England, the United States, and Sardinia.

Wulfenite, $PbMoO_4$: tetragonal system; sg 6.78; hardness 2.7–3.

Its crystals are usually tabular or equidimensional, coloured waxy yellow, orange yellow, orange red, reddish brown, with a resinous or adamantine lustre. It forms in alteration areas of lead-zinc deposits. It is fairly well known in Bleiberg, Carinthia (Austria); M'Fouati in the Congo; Sierra de los Lamentos in the state of Chihuahua (Mexico). In the United States fine yellow and reddish-orange crystals are found at Wheatley's Mine, near Phoenixville, Pennsylvania, and good crystals come from the Organ Mountains, near Las Cruces, New Mexico.

Tabular wulfenite crystal. Red Cloud Mine (Arizona)

Class VII: phosphates, arsenates, and vanadates

This class contains a considerable number of minerals, many of which are most attractive with their brilliant colours, although they are little known. They are characterized, in the anionic group, by the presence of phosphorus in the *phosphates*, arsenic in the *arsenates*, and vanadium in the *vanadates*.

They are almost always secondary minerals, that is, they have formed at the expense of others. Some of them are extremely useful in that they provide rare chemical elements for extraction.

Lithiophilite, xenotime, monazite

Lithiophilite, Li,Mn,FePO$_4$, is a useful lithium mineral; orthorhombic, generally compact or in malformed crystals coloured brownish yellow with a resinous lustre. It is a rare mineral found in the pegmatites of Mangualde in Portugal, Varuträsk in Sweden, and Custer in South Dakota.

Interesting rare earth minerals are: *xenotime*, YPO$_4$, tetragonal, and *monazite*, CePO$_4$, monoclinic. Both are typical of pegmatites and volcanic rocks, have yellowish-brown prisms, and can be found in Scandinavia and Madagascar.

Libethenite, olivenite, adamite

Libethenite, Cu$_2$(OH)PO$_4$, is a secondary mineral in copper deposits; it can occur in small, thin, orthorhombic prisms, coloured dark olive green; it is translucent, with a vitreous lustre. It can be seen in, among other places, Cornwall (England), Lubietová (Czechoslovakia), and Chuquicamata (Chile).

Very similar to libethenite, with which it is often associated, is *olivenite*, Cu$_2$(OH)AsO$_4$, found in small acicular or prismatic crystals, sometimes in globular, radiating fibrous aggregates, olive green or greenish brown. It can be obtained in various parts of Cornwall

Right: Libethenite crystals. Rheinbreitbach, Siebengebirge (Germany)

Far right: Olivenite. Cornwall (England)

Far left: Close association of adamite crystals. Ojuela Mine, Mapimi (Mexico)

Left: Cornetite. Empire Nevada Mine, Yerington (Nevada)

and Cumberland (England), Tsumeb (South West Africa), Cape Garonne in Var (France), and Tintic in Utah.

Often associated with the two previous minerals is *adamite*, $Zn_2(OH)AsO_4$, which forms aggregates of small, orthorhombic, many-faced crystals, or radiating aggregates covering fairly extensive surfaces. The vitreous lustre is complemented by the delicate yellow-green colour in a variety of tones, or emerald green. It is sometimes fluorescent in lemon yellow under ultra-violet light. The most noteworthy deposits are at Lavrion (Greece), Tsumeb (South West Africa), Ojuela Mine near Mapimí, Durango (Mexico).

Cornetite, descloizite, brazilianite, crandallite

An example of a rare copper phosphate is *cornetite*, $Cu_3(OH)_3PO_4$, in aggregates of small dark-blue crystals, found in Nevada and the Congo.

Descloizite, $Pb(Zn,Cu)VO_4OH$, is orthorhombic and is found in aggregates or formations which are more or less parallel, consisting of crystals in irregular groups with a clearly pyramidal shape. The colour varies from dark brown to reddish. *Mottramite* is also found with it and is of similar composition but with a higher copper content. The best-known places of origin are at Tsumeb and Grootfontein in South West Africa, and Broken Hill in Zambia. It is also found in New Mexico, and Arizona.

Growth of descloizite crystals. Tsumeb (South West Africa)

Right: Semi-transparent
individual crystal of
brazilianite. Brazil

Brazilianite, $NaAl_3(PO_4)_2(OH)_4$, monoclinic; this
mineral was discovered in 1945 at Conselheiro Pena,
Minas Gerais (Brazil), in good prismatic crystals of a
yellow-green colour, translucent or transparent. The
excellent quality and abundance of this mineral have
made its discovery an exceptional event in the recent
history of mineralogy, because any new varieties
discovered now are usually found in small quantities
and in poor quality crystals.

Crandallite, a phosphate of calcium and aluminium,
is a rare mineral in fibrous radiating forms, opaque,
whitish, obtainable in Bavaria (Germany), Belgium,
and Utah.

Apatite, $Ca_5F(PO_4)_3$: hexagonal system; sg 3.1–3.2;
hardness 5.

This is a fairly common mineral found in well-shaped
crystals, prismatic, green gray, asparagus green, sea
green, yellow brown, pink, violet, milk white, or
colourless; transparent, translucent, or opaque, with a
greasy lustre. It is found in pegmatites, in metamorphic
and igneous rocks, and in ore veins. Yellowish-green
apatite is quite well known in the magnetite deposits of
Cerro de Mercado, Durango (Mexico), and the violet
variety at Ehrenfriedersdorf in Saxony (Germany). At
Hammond, New York, large crystals occur; yellow

Below: White tabular
apatite with chlorite.
Valle Aurina, Bolzano
(Italy)

67

crystals are found in Alexander County, North Carolina, and at very many other localities.

Pyromorphite, mimetite, vanadinite

Pyromorphite, $Pb_5Cl(PO_4)_3$, has small, prismatic, hexagonal crystals coloured bright green, yellow, brown, gray, with a vitreous or greasy lustre; like mimetite and vanadinite, it is a secondary mineral found in many lead deposits. The best examples have been obtained in Germany and Broken Hill, Australia. It is found in the Wheatley Mine in Pennsylvania; in Davidson County, North Carolina; the Coeur d'Alene district, and at Mace, Mullan, Wardner, and Burke, all in Idaho.

Mimetite, $Pb_5Cl(AsO_4)_3$, is often very similar to pyromorphite, but its colour tends more towards yellow and brown; in the *campylite* variety the prismatic crystals show curved faces and are rounded (melon-shaped): good yellow-brown crystals on white quartz are obtained at Dry Gill in Cumberland, England. Relatively large mimetite crystals come from various German localities. In the United States it is found in the Wheatley Mine, and at Eureka in the Tintic district, Utah.

Vanadinite, $Pb_5Cl(VO_4)_3$, occurs in aggregates of small hexagonal crystals coloured orange, ruby red, reddish brown, often tabular or equidimensional. Very famous specimens have been found in the

Tabular crystals of violet apatite. Ehrenfriedersdorf, Saxony (Germany)

Left: Pyromorphite in small prismatic crystals. Mottarone, Lake Maggiore (Italy)

Right: Mimetite. Johanngeorgenstadt, Saxony (Germany)

68

Translucent vivianite. Llallagua, Uncía (Bolivia). Vivianite, showing a clear pleochroism, often forms in lacustrine areas as a result of the chemical action of iron salts with phosphatic organic substances

Below left: Vanadinite. Old Yuma Mine (Arizona)

Below: Smooth surface of a variscite nodule. Fairfield (Utah)

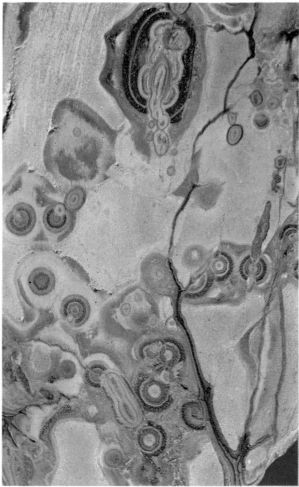

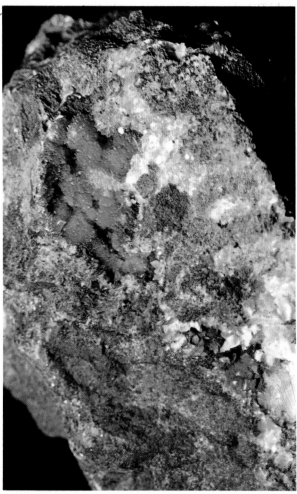

Apache Mine and Old Yuma Mine in Arizona, and also in parts of New Mexico. Splendid crystals are found at Mibladen, Morocco, and it comes from Serra de Cordoba, Argentina, and Ottoshoop, Marico district, Transvaal.

Variscite, hopeite, vivianite

Variscite, $AlPO_4.2H_2O$, is usually present in nodules, veins, or irregular masses of an apple-green or leek-green colour, with a porcellaneous lustre. The best examples, sometimes used for ornamental purposes, come from Utah.

Hopeite, $Zn_3(PO_4)_2.4H_2O$, is a secondary mineral of zinc deposits; it occurs in crusts of tabular, reddish-brown crystals found in the much-mined Broken Hill in Zambia.

Vivianite, $Fe_3(PO_4)_2.8H_2O$, is a secondary mineral in a great many ore veins, and occurs in monoclinic crystals which sometimes have rounded faces, or otherwise in patinas and earthy masses, dark blue or dark green. Transparent or translucent, easily crumbled, it has a vitreous or pearly lustre, low specific gravity, and a very low degree of hardness. Good crystals are found at Poopó in Bolivia and Trepča in Yugoslavia. An exceptional deposit in lacustrine clay was found a few years ago near Anloua in Cameroun, which consisted of isolated crystals or groups of several individuals more than a metre long.

Left: Euchroite in minute green crystals. Lubietová (Czechoslovakia)

Right: Wavellite in fibrous-radiating rosettes. Barnstaple, Devonshire (England)

Above: Autunite in plates arranged in a fan formation. Peveragno, Piedmont (Italy). Like the other uranium 'micas', it has a more dehydrated 'meta' phase which, however, is reversible. Almost all the examples in the collection are meta phase (meta autunite, meta torbernite, etc)

Right: Equidimensional, pseudo-cubic torbernite crystals. Cornwall (England)

Destinezite, wavellite, turquoise

Destinezite is a hydrous sulphate-phosphate of iron, which forms yellowish-white earthy nodules; it occurs in Thüringen (Germany), Finistère (France), and Bohemia (Czechoslovakia).

Wavellite, $Al_6(PO_4)_4(OH)_6.9H_2O$, forms stalactitic masses, or well-shaped aggregates of globules, or rosettes of radiating acicular individuals; the lustre is vitreous, the colour greenish white or yellowish. The main places where it has been found are Bohemia (Czechoslovakia), Romania, Cornwall (England), and Bolivia. In the United States wavellite occurs in a number of localities in Pennsylvania and near Avant, Arkansas.

Turquoise, $CuAl_6(PO_4)_4(OH)_8.5H_2O$, is a fairly valuable mineral, used as a gemstone because of its beautiful sky-blue colour. It is very occasionally found in prismatic triclinic crystals, more frequently in compact masses or veins with a hardness of 5–6 and a low specific gravity (2.6). It is of secondary origin and as such it fills cavities in brecciated rocks which are usually volcanic in origin, such as the trachytes of Nishapur in Khorasan (Iran), which provide the best stones. Other localities are Siberia, Turkestan, New Mexico, and Arizona.

Erythrite, annabergite, pharmacolite, euchroite

Erythrite, or 'cobalt bloom', $Co_3(AsO_4)_2.8H_2O$, owes its name to the dark-red colour of its tabular or lamellar crystals, which are in fibrous, silky, or earthy aggregates formed by alteration of various cobalt minerals. Good examples are found in Schneeberg in Saxony (Germany), and Bou Azzer in Morocco. It is found at Cobalt, Ontario, in pinkish crusts and small crystals. Good specimens are not common in the United States, although it has been noted in Nevada, Arizona, New Mexico, and California.

Cabrerite is a magnesian variety of *annabergite*, or 'nickel bloom', $Ni_3(AsO_4)_2.8H_2O$, which is found in aggregates of apple-green lamellar crystals. It is a rare alteration product of nickel sulphides, found in Saxony, Ontario, and Lavrion (Greece). The best United States occurrence is in Humboldt County, Nevada.

Pharmacolite, $CaHAsO_4.2H_2O$, occurs in tufts of acicular or fibrous crystals which are white and have a silky lustre; the cobalt veins of Germany are the most usual places where it is found.

Euchroite, a basic hydrous arsenate of copper, has small emerald-green crystals. It is found with other secondary minerals in oxidized parts of cuprite deposits.

Tyrolite, liroconite, pharmacosiderite

Tyrolite is a basic hydrous arsenate of calcium and copper, crystallized in the orthorhombic system, which forms in the oxidation areas of some deposits of copper with erythrite, malachite, azurite, chrysocolla, etc.

It is usually found in scaly patinas or in lamellar aggregates coloured apple green, gray green tending towards blue, with a pearly lustre on the cleavage surface. It occurs in Austria, Czechoslovakia, Germany, France, and at the Mammoth Mine, in the Tintic district of Utah in the United States.

Liroconite is a monoclinic, basic hydrous arsenate of copper and aluminium, and it, too, is typical of oxidation areas in copper deposits. It forms deep sky-blue crystals of an approximately octahedral form. The best-known areas where it is found are Cornwall and Devon in England; also Germany and the Urals.

Pharmacosiderite is a basic hydrous arsenate of potassium and iron, crystallized in the isometric system, generally in cubes. The colour is usually olive green or yellowish brown, but it can also be hyacinth pink, violet, or emerald green; the lustre is greasy or adamantine. It is usually a secondary mineral from the alteration of arsenates. Good crystals are found in several parts of Germany, especially Schwarzenberg in Saxony, also Cornwall (England), Czechoslovakia, France, and at the Mammoth Mine in Utah.

Uranium phosphates

Several secondary minerals of uranium are classified as phosphates; they are usually various shades of yellow or green and often accompany pitchblende. A few of these are noteworthy.

Torbernite, $Cu(UO_2)_2(PO_4)_2.12H_2O$: orthorhombic system; sg 3.22; hardness 2.2. Also known as *chalcolite*, it almost always forms aggregates of tabular or lamellar crystals with clear-cut angles, grass green or emerald green. It is not fluorescent in ultraviolet light and it has a vitreous lustre. It is found at Redruth in Cornwall (England), Shinkolobwe in Katanga, at Mount Painter, Flinders Range, Australia, and in France.

Autunite, $Ca(UO_2)_2(PO_4)_2.12H_2O$: orthorhombic system; sg 3.1; hardness 2–2.5. An extremely well-known uranium mineral, it has small, lamellar, flattened crystals, or forms semi-parallel growths arranged in a fan-like shape; the colour ranges from lemon yellow to pale green. The green fluorescence which it always shows in ultraviolet light makes it possible to distinguish it from torbernite. The best examples come from Daybreak Mine, Mount Spokane, Washington State; very attractive yellow ones can be found at Margnac, Haute Vienne (France). Specimens similar to those on Mount Spokane can be found in Italy at Peveragno (Piedmont), where yellow and pale green autunite is also common. It is also found in northern Portugal, at Sabugal, south east of Guarda, and at Vizeu.

Uranocircite, $Ba(UO_2)_2(PO_4)_2.8H_2O$, is much rarer and its square yellowish-green crystals make it resemble autunite. It comes from Germany.

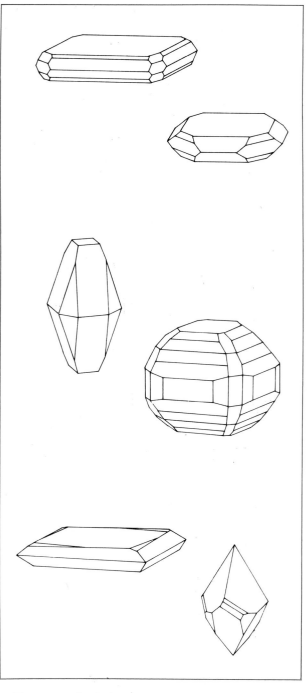

Torbernite. Top: The most common lamellar forms similar to the other uranium micas Centre: Two rarer forms – bipyramidal and blunt-ended. Bottom: Wulfenite, the common lamellar form and the rarer bipyramidal one

Zeunerite, $Cu(UO_2)_2(AsO_4)_2.16H_2O$, is very much like torbernite in form and colour but it is much rarer. It is found in the United States, France, Katanga, and Italy.

Phosphuranylite, $Ca(UO_2)_4(OH)_4(PO_4)_2.7H_2O$, usually occurs in patinas and crusts coloured yellow gold, in uraninite pegmatites. It can be found in various localities of the United States, Portugal, and France.

Class VIII: silicates

This group includes about one third of all known minerals; several are used as gems, others serve industrial purposes; some are very common components of rocks, others are extremely rare and much sought-after. Most of them are very hard (6–8), and they can rarely be altered.

They are characterized by the structural group SiO_4, composed of a central silicon atom placed in the middle of four oxygen atoms arranged in a tetrahedron. In the structure of a silicate these tetrahedra can be isolated from each other *(nesosilicates)*, or grouped in twos *(sorosilicates)*, or linked together in rings *(cyclosilicates)*, in very elongated chains *(inosilicates)*, or arranged in flat layers *(phillosilicates)*, or, finally, in three-dimensional structures *(tektosilicates)*.

A characteristic which cannot often be found in minerals of the other classes is the presence of the *isomorphous* series, that is the chemical composition varies gradually from one mineral to another so that the first and last are entirely different from each other. For example, in the case of olivine, between the beginning of the series which is rich in magnesium (forsterite) and the opposite extreme which is rich in iron (fayalite) there is a wide range of minerals with an intermediate composition.

Nesosilicates

Olivine, $(Mg,Fe)_2SiO_4$: orthorhombic system; sg 3.27–3.37; hardness 6.5–7.

Olivine, also called *chrysolite* or *peridot*, is in fact made up of a series varying from *forsterite*, Mg_2SiO_4, to *fayalite*, Fe_2SiO_4. It only occasionally occurs in separate crystals, more often in small granular masses in igneous rocks with a low silica content; the colour varies from gray green to olive green and yellowish brown. The large transparent crystals used for

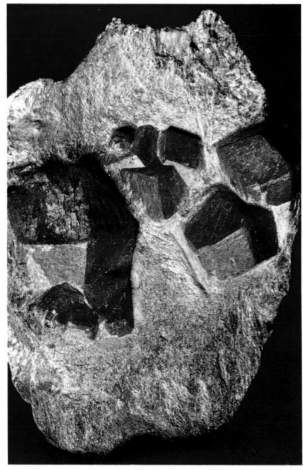

Right: Prismatic olivine crystals. Mt Somma, Naples (Italy)

Far right: Almandine garnet in rhombohedra. Oetztal (Austria)

Below: Some commoner habits of isometric minerals such as garnet, leucite, analcime. Above left: Rhombododecahedron, typical form for garnet. Above centre: Icositetrahedron or leucitohedron, typical form for leucite and analcime. Above right and below: Various crystal habits of garnet

jewellery are called *chrysolite*, and can be obtained in the island of St John in the Red Sea. It occurs in boulders of basalt at Thetford, Vermont.

The garnet group

Garnets, which belong to the isometric system, are very common minerals in igneous and metamorphic rocks. Their habit is usually rhombododecahedral or icositetrahedral, though dimensions and colour vary.

Pyrope, $Mg_3Al_2(SiO_4)_3$, is bright red and often transparent, and has a specific gravity of 3.51. It is quite rare and is used for ornamental purposes; it is found in Bohemia at Měrunice, in Switzerland at Gorduno near Bellinzona, in South Africa in the Kimberley and other diamond mines, and in transparent grains in Arizona and New Mexico.

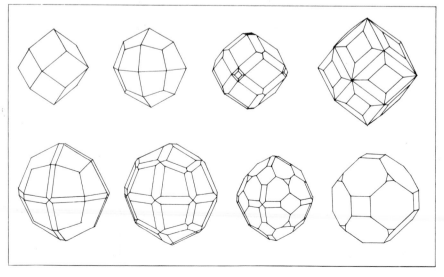

Right: Hessonite, a variety of grossular garnet, with chlorite. Val d'Ala, Piedmont (Italy)

Above: Uvarovite. Urals (USSR)

Below: Demantoid, a variety of andradite garnet, with asbestos. Val Malenco, Sondrio (Italy)

Below: Zircon. Slatoust, Urals (USSR)

Almandine, $Fe_3Al_2(SiO_4)_3$, is a very common garnet coloured wine red, brown red, sometimes transparent, but more often translucent. It can be distinguished from pyrope by its higher specific gravity (4.25). It occurs widely in the Alps, and sometimes it can be the size of a fist or even larger. Large crystals are found in the Adirondacks at North Creek, New York, where it is mined for garnet paper.

Spessartite, $Mn_3Al_2(SiO_4)_3$, is an orange or yellow-orange garnet which is fairly uncommon and is found in Virginia and California.

Grossularite, $Ca_3Al_2(SiO_4)_3$. The most common colours of this garnet vary from orange to reddish yellow, yellowish green to colourless. *Hessonite* is the ferrous variety which is found, for example, in Val d'Ala, Piedmont (Italy). Large pink crystals are obtained at Morelos and Chihuahua in Mexico.

Andradite, $Ca_3Fe_2(SiO_4)_3$. Its colour ranges from a warm yellow in the *topazolite* variety, to grass green and emerald green in the *demantoid* variety, to black in the titaniferous variety called *melanite*. The first occurs in Val d'Ala, Piedmont (Italy) and Nizhniy-Tagil (Urals), the last in volcanic tuff in central Italy.

Uvarovite, $Ca_3Cr_2(SiO_4)_3$. This is a rare, dark-green garnet found in the Urals and in Finland.

Zircon, $Zr(SiO_4)$: tetragonal system; sg 4.6–4.7; hardness 7.5.

Prismatic crystals of Andalusite. Tyrol (Austria)

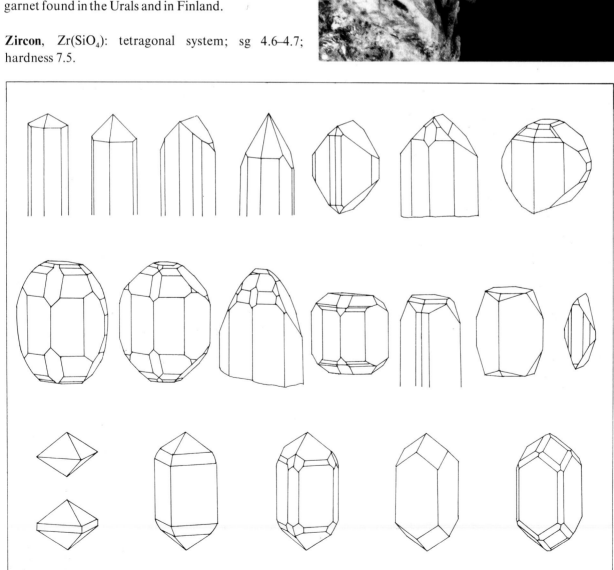

Topaz. Although this mineral has many types of crystal face it normally develops a prismatic form and normal lengthwise cleavage. Bottom row: Zircon, the first two from the left with a bipyramidal habit, the others with a prismatic habit

It is a common accessory of volcanic rock, where it usually occurs in prismatic crystals; it takes the form of rounded granules when in alluvial deposits. It is either colourless or it varies from dark brown to yellow brown or yellowish green. The lustre is adamantine and very bright; the precious varieties such as *hyacinth*, which is orange or reddish, are normally transparent. The blue colour seems to have been caused by heat. *Malacon* is an altered zircon. Those found in Ceylon and Thailand are used as gems, while good crystals are found in the Urals, Australia, Sweden, and Madagascar. Several different types of crystals are found in Henderson County, North Carolina; Sparta, New Jersey; Natural Bridge, New York; Princetown, Pennsylvania; Renfrew, Ontario; and Cheyenne Mountain near Colorado Springs, Colorado.

Sillimanite, andalusite, kyanite

These are three silicates of aluminium characteristic of metamorphic rocks.

Sillimanite, Al_2SiO_5, is usually found in fibrous white silky aggregates, or otherwise in malformed greenish-gray crystals. It is widely disseminated but unspectacular. It can be found at Castione (Bellinzona), Ossola (Piedmont), Şondalo (Valtellina), Sila (Calabria), all in Italy, or at Bodenmais in Bavaria (Germany). It occurs in fibrous-embedded masses at Worcester, Massachusetts, and Norwich and Willimantic, Connecticut; also in New York and Pennsylvania.

Andalusite, Al_2SiO_5, is almost always in blunt, orthorhombic, prismatic crystals of an iron-gray colour, or otherwise in radiating aggregates of acicular pink crystals. Very beautiful groups are found near Innsbruck in the Tyrol. The variety *chiastolite* shows in a transverse section a dark cross shape caused by carbon particles. Various localities in Spain provide chiastolite, which is used as a good-luck stone.

Kyanite, or *disthene*, Al_2SiO_5, occurs in flat, sometimes very long crystals, in bundles, radiating or isolated, coloured sky blue or blue, with a vitreous lustre, often transparent or translucent. Very beautiful and famous groups of blue kyanite and brown staurolite are found on white schist in Pizzo Forno (Ticino Canton, Switzerland). *Rhaetizite* is the gray or colourless variety found in Alto Adige (Italy), and the Tyrol (Austria).

Topaz, $Al_2SiO_4(OH,F)_2$: orthorhombic system; sg 3.4–3.6; hardness 8.

This mineral has been used as a gem from the earliest times, especially in its wine-yellow and amber varieties. It has blunt-ended prismatic crystals which can sometimes be enormous; they are colourless, blue, reddish, pink, and usually transparent. The most highly valued gem stones are found in Ouro Prêto

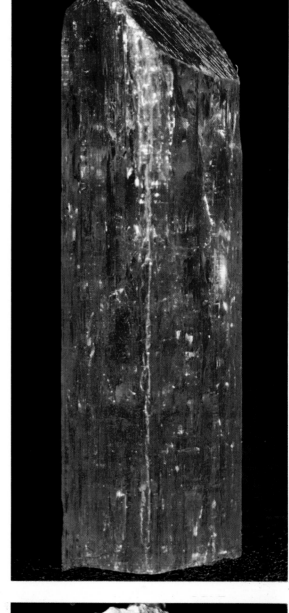

Topaz, prismatic crystal. Brazil

Kyanite with brown staurolite on paragonitic schist. Pizzo Forno, Ticino (Switzerland)

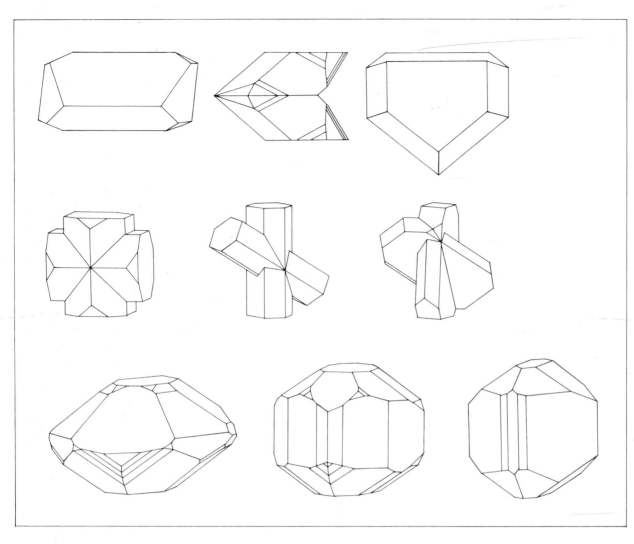

in the state of Minas Gerais (Brazil); blue transparent hyaline crystals are also obtained in Brazil. Excellent examples can also be found in the Nerchinsk Range, Transbaikalia (Siberia), and Schneckenstein, Saxony (Germany). In the United States fine crystals come from Streeter, Mason County, Texas; transparent crystals are found in rhyolite in the Thomas Mountains, Utah, while in Colorado fine colourless or pale-blue crystals come from the Pikes Peak and Crystal Peak.

Staurolite, $(Fe,Mg)_4Al_{18}Si_8O_{46}(OH)_2$: orthorhombic system; sg 3.65–3.77; hardness 7–7.5.

Its name derives from the Greek word *staurós*, meaning 'cross', in allusion to its cruciform twins. The crystals are blunt ended and prismatic, or long and flat; the colour varies from dark gray to reddish brown; the lustre is vitreous. Sometimes isolated crystals are covered with an earthy pàtina due to alteration. Interesting cruciform twins come from Pilar in New Mexico and Morbihan in France. Elegant specimens of blue kyanite on white schist with brown staurolite come from Pizzo Forno in the Ticino Canton (Switzerland).

Humite, titanclinohumite

Very similar to staurolite, in its chemical composition, is *humite*, $Mg_7Si_3O_{12}(F,OH)_2$, only occasionally

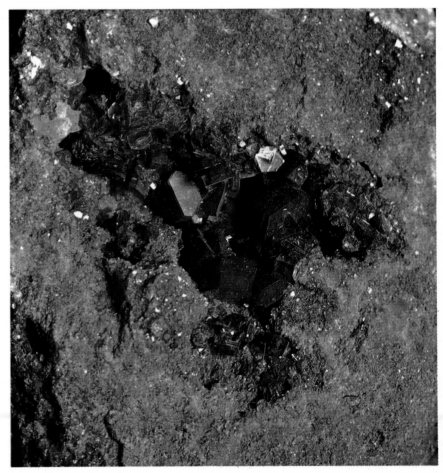

*Right: Titanclino-
humite, also called
titanolivine, in brown-
red nodules. Val
Malenco, Sondrio
(Italy)*

found as separate crystals coloured yellow brown, such as the ones from Mount Somma, and altered ones at Tilly Forster Mine, New York.

The humite family includes *titanclinohumite*, also known as *titanolivine*, $(Mg,Ti)_9Si_4O_{16}(OH)_2$, which is found in compact dark reddish-brown masses in the serpentinites of the Alps and the Ligurian Apennines.

Braunite

Braunite, $Mn''Mn'''_6SiO_{12}$, is a useful manganese mineral rarely found in prismatic crystals, but more often in compact granular masses coloured black or iron gray. It is frequently associated with other manganese minerals and can be found in the Aosta Valley, Italy, Långbanshyttan in Sweden, and at Kacharwaki, Nagpur (India).

Titanite, $CaTiSiO_5$: monoclinic system; sg 3.4–3.56; hardness 5–5.5.

It is a common mineral of igneous and metamorphic rocks, as in the Alps, where it occurs in wedge-shaped tabular crystals. These crystals are often twinned either by contact or by co-penetration. The colour varies from white to yellow or green; very occasionally it is red, and this manganese variety, called *greenovite*, is found in the Aosta Valley; sometimes it is brownish red, like the prismatic titanite of Khibinskaya Tundra in the Kola Peninsula (USSR). Very lovely crystals have been obtained at Ossola in Piedmont, Ticino Canton

*Right: Titanite in
wedge-shaped twinned
crystals. St Gotthard
(Switzerland)*

and in Graubünden in Switzerland, Alto Adige, and the Tyrol, and these crystals are often covered with green pulverulent chlorite.

Datolite, dumortierite, howlite, grandidierite

Datolite, $CaBSiO_4OH$, is a monoclinic mineral usually associated with volcanic zeolites and cuprite deposits; its crystals are many faced with a high vitreous lustre, colourless, white, or pale green.

Good specimens are mined in Prospect Park near Paterson, New Jersey, and at Keweenaw, Michigan, while large crystals occur near Hartford and at Tariffville in Connecticut. Datolite is also obtained from Montagu County, Tasmania, (Australia), and from Arendal in Norway.

Dumortierite, $(Al,Fe)_7BSi_3O_{18}$, is a rare borate mineral found in compact masses or fibrous aggregates of a dark-blue colour; although fairly common in

Below: Compact dumortierite. Madagascar. This mineral is also used for ornamental purposes

Madagascar, it is most frequently encountered in the west of the United States, and has been mined in Oreana, Nevada, for spark plug ceramics. The blue variety is found in Los Angeles County, California, and New York City building excavations produce fair dumortierite needles.

Howlite, $Ca_2B_5SiO_9(OH)_5$, is also a rare borate mineral which occurs in irregular, whitish reniform masses, in the borate deposits of California.

Above: Ilvaite. Rio Marino, Elba (Italy)

Grandidierite, $(Mg,Fe)Al_3BSiO_9$, orthorhombic, is another borate mineral though typical of some pegmatites; it is found in aggregates of prismatic crystals coloured greenish blue, in Port Dauphin, southern Madagascar.

Gadolinite, uranophane

Gadolinite, $Be_2FeY_2Si_2O_{10}$, is a rare earth mineral which has prismatic monoclinic crystals or compact masses, coloured pitch black with a conchoidal fracture. The best examples come from Hiterö and Iveland in Norway, and Ytterby in Sweden. In the United States it is found in Llano County, Texas, and on the west bank of the Colorado River, near Bluffton.

Uranophane or *uranotile*, $Ca(UO_2)_2Si_2O_7 \cdot 6H_2O$, presents clusters of bright yellow acicular crystals which can also be in yellowish patinas. It has a green fluorescence in ultraviolet light. Excellent examples come from Ambrosia, New Mexico.

Sorosilicates

Left: Uranophane (or uranotile) in acicular crystals. Ambrosia (New Mexico)

Gehlenite, melilite, cuspidine

Gehlenite, $Ca_2Al_2SiO_7$, is a rare mineral which forms in calcareous rocks metamorphosed by igneous intrusions; it has short, whitish, tetragonal prisms, similar to those found in Rif Taramelli in the Monzoni hills, Trento (Italy).

Present in many volcanic rocks is *melilite*, which occurs in short tetragonal crystals, whitish, yellow, or reddish, in leucite-nepheline basalt in central Italy, in the area round Mount Vesuvius, the Rhineland (Germany), and Hawaii among other places.

Another volcanic product of the Latium and Vesuvius regions in Italy is *cuspidine*, $Ca_4Si_2O_7(F,OH)_2$, found in pointed crystals that can be whitish or reddish.

Ilvaite, $CaFe''_2Fe'''Si_2O_8OH$: orthorhombic system; sg 3.99–4.05; hardness 5.5–6.

It was first found on the island of Elba – *Ilva* in Latin – where it is found in compact masses, aggregates of acicular individuals or prismatic crystals striated longitudinally, coloured pitch black or red due to incipient oxidation, with a semi-metallic or resinous lustre. Crystals with the best shape and lustre are found in Rio Marina and Cape Calamita on Elba. It can also be found in fine crystals at the South Mountain Mine, Owyhee County, Idaho,

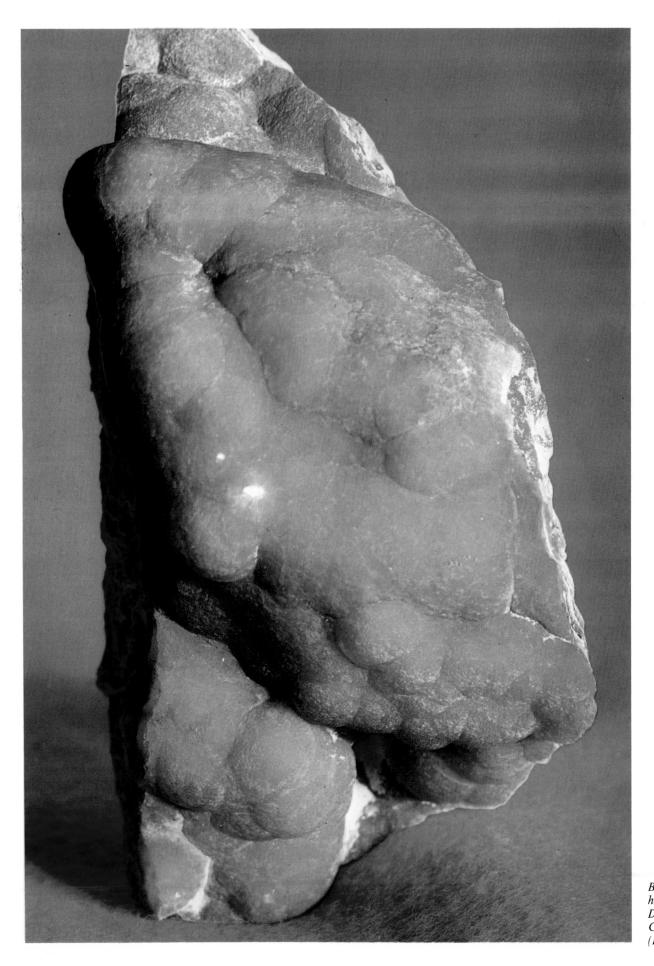

Blue mammillary hemimorphite. Sa Duchessa Mine, Cagliari, Sardinia (Italy)

and large opaque crystals are found at Serifos in the Cyclades (Aegean Sea).

Hemimorphite, $Zn_4Si_2O_7(OH)_2.H_2O$: orthorhombic system; sg 3.40–3.50; hardness 4.5–5.

In miners' jargon both smithsonite and hemimorphite are known by the generic name of 'calamine', although this term is more commonly restricted to hemimorphite. It is a secondary mineral of zinc deposits where it forms stalactitic, concretionary, mammillated and fibrous masses, or, more rarely, small prismatic crystals. The colour varies from milk white to brownish yellow or sea blue. The best examples are white crystals as much as 5 centimetres (2 inches) long on a yellow limonitic matrix, which come from the Ojuela Mine near Mapimí, Durango (Mexico). Concretionary and mammillary crusts, coloured intense blue, are found in the Sa Duchessa Mine near Domusnovas, Cagliari (Sardinia), where hemimorphite accompanies chrysocolla, bronchantite, malachite, cuprite, etc. In the United States it has been extensively worked at Friedensville, Lehigh County, Pennsylvania, and fine crystals can be found at Leadville, Lake County, Colorado; the Organ Mountains, New Mexico; and Elkhorn, Jefferson County, Montana.

Epidote, $Ca_2(Al,Fe)_3Si_3O_{12}OH$: monoclinic system; sg 3.4–3.5; hardness 6–7.

This mineral is one of a series including several varieties, crystallized in the monoclinic or orthorhombic systems and fairly common as an accessory of various rocks. The true epidote has prismatic crystals of a colour varying from dark green to brown green, pistachio green (*pistacite* variety), and pale green. Splendid groups of fine dark-green and brown-green crystals several inches long are obtained at Knappenwand in Untersulzbachtal, Austria, and in several mountain localities such as Oisans in Dauphiné (France), and Val d'Ossola in Piedmont (Italy). There are also fine large crystals in the United States at Haddam, Middlesex County, Connecticut, near Salida in Colorado, and at Riverside, California.

A manganese epidote is called *piedmontite*, which is found in clustered aggregates of acicular individuals coloured violet brown, occurring in the manganese deposits of San Marcel in the Aosta Valley (Italy).

Zoisite is orthorhombic and *clinozoisite* monoclinic,

and they both belong to the above family, usually forming granular or acicular aggregates coloured greenish, gray, pink (*thulite* variety). Zoisite is found at Goshen, Hampshire County, and Conway, Franklin County, both in Massachusetts, and also at Leiperville, Delaware County, Pennsylvania. *Allanite*, or *orthite*, is an epidote containing rare earths, it is pitch black and has a conchoidal fracture.

Idocrase, $Ca_{10}(Mg,Fe)_2Al_4Si_9O_{34}(OH)_4$: tetragonal system; sg 3.35–3.47; hardness 6.5.

This mineral, also called *vesuvianite*, is named after

Right: Blue benitoite with black neptunite in white natrolite. San Benito County (California)

Far right: Drawings of idocrase; the first three are the most common prismatic shapes; the others are less common tabular forms

Left: Idocrase. Chigi Park, Rome (Italy)

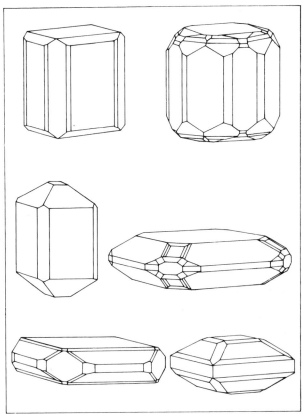

Mount Vesuvius where it was first found towards the end of the eighteenth century. It has prismatic, rather stubby, crystals or otherwise it forms compact masses whose colour ranges from grass green to brown green, brown, or black; its lustre is vitreous; it is sometimes transparent but more usually translucent or opaque. Some good crystals are mined in Quebec, and in the river Vilyui in Siberia.

In the United States it is found at Auburn and Sanford, Maine; there are also grayish and yellow-brown crystals near Amity, New York, Franklin, New Jersey, and Magnet Cove, near Hot Springs, Garland County, Arkansas.

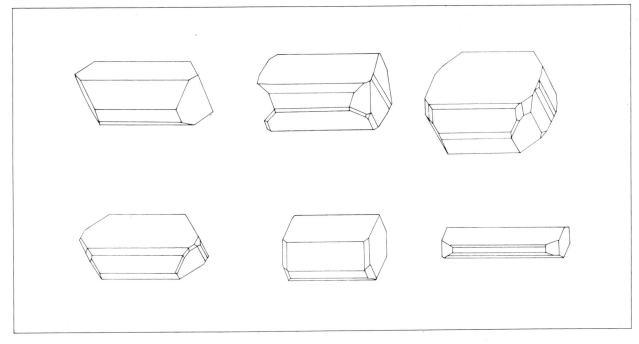

Right: Epidote, the most common prismatic shapes

Eudialyte. Khibinskaya Tundra, Kola Peninsula (USSR)

Cyclosilicates

Benitoite, eudialyte, taramellite, axinite

Benitoite, $BaTiSi_3O_9$. It is named after San Benito County in California, the only place where this mineral can be found. The crystals are stumpy, prismatic, and coloured sky blue; they are found in association with black neptunite, in white natrolite veins in altered serpentinite rocks.

Eudialyte, $Na_4(CaFe)_2ZrSi_6O_{17}(OH,Cl)_2$. This is a syenite pegmatite mineral found in granules and irregular masses coloured brick red, pink, or brown. The best specimens come from Magnet Cove in Arkansas, and from Greenland, Norway, and the Kola peninsula in the Soviet Union.

Taramellite is a rare silicate of barium and iron, and is found in fibrous aggregates of a dark-violet

Tinzenite, manganiferous variety of axinite with a fibrous appearance. Cassagna, Genoa (Italy)

colour in marble at Candoglia in Val d'Ossola (Piedmont).

Axinite, $Ca_2(Mn,Fe)Al_2BSi_4O_{15}OH$. It has triclinic crystals with a lamellar appearance, wedge-shaped, coloured brown, violet brown, gray brown, with a vitreous lustre and good transparency. Famous localities are Bourg d'Oisans in Dauphiné (France), Coarsegold, Madera County, in California, and Scopi in Graubünden (Switzerland). It is also found in Cornwall, England; Franklin, New Jersey (as honey-coloured crystals); and San Diego County, California. The magnesitic variety, *tinzenite*, is lamellar, radiating, coloured orange, reddish, sulphur yellow, and is named after the Tinzen deposit in Switzerland, although it is also found in the manganese mines of Cassagna and Gambatesa near Chiavari in Liguria (Italy).

Axinite. Le Bourg d'Oisans, Dauphiné (France). This mineral is associated with hydrothermal deposits and is found in various parts of the Alps

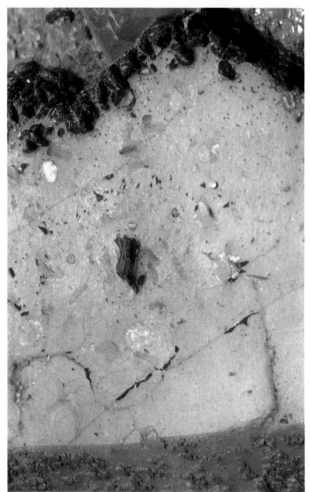

Beryl, $Be_3Al_2Si_6O_{18}$: hexagonal system; sg 2.63–2.80; hardness 7.5–8.

This mineral has long been appreciated and sought-after as a gemstone, especially in its green transparent variety *(emerald)*, and blue *(aquamarine)*. It appears in prisms with a hexagonal section, the size of which varies greatly; its colour can be intense green, blue, azure, yellow *(heliodor)*, dark rose *(morganite)*, grayish white; it can be transparent, translucent, or opaque. Besides its use in jewellery, beryllium is also extracted from it. It is a typical pegmatite mineral, and large deposits exist in Brazil, in the state of Minas Gerais; in Madagascar, and in the Urals. Very famous emeralds are obtained from Muzo in Colombia; in Europe there is the old emerald mine of Habachtal in the Salzburg region in Austria. In the United States very large crystals have been found at Albany, Maine, and Sullivan County, New Hampshire.

Long prismatic crystals typical of aquamarine and heliodore, for example, compared with shorter beryl crystals, which are rich in alkalis

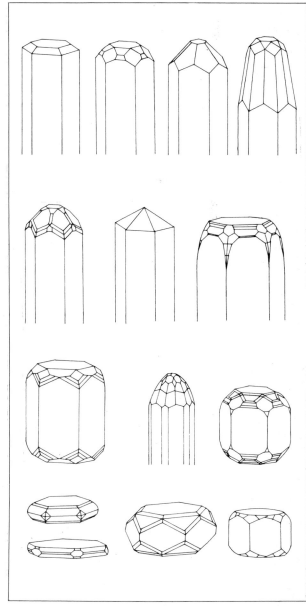

Black tourmaline (schörlite) in pegmatite. Olgiasca, Lake Como (Italy)

The rare *bazzite* is a silicate of scandium, beryllium, and iron, and consists of sky-blue prisms. It is found at Baveno on Lake Maggiore in Italy, and also in various parts of Switzerland and the Soviet Union.

Cordierite

Cordierite, $(Mg,Fe)_2Al_4Si_5O_{18}$, is also known as *dichroite*. It crystallizes in the orthorhombic system and these crystals are short pseudo-hexagonal prisms which can be colourless, blue, or violet blue. The transparent violet-blue variety, known as *iolite*, is used in jewellery. If it is in granules or fragmented crystals it can easily be mistaken for quartz. Good crystals are found at Bodenmais in Bavaria (Germany), and Orijärvi in Finland. In the United States it is found chiefly in Connecticut and New Hampshire.

Tourmaline, $(Na,Ca)(Li,Mg,Fe^{..},Al)_3(Al,Fe^{...})_6B_3Si_6O_{27}(O,OH,F)_4$: trigonal system; sg 2.98–3.20; hardness 7.75.

The crystals have a prismatic, pyramidal habit, vertically striated. The above formula applies to the most common black variety called *schorl*. Crystals of other colours contain lithium, magnesium, and calcium; these colours can be pink or red *(elbaite,*

Right: Pink prismatic tourmaline crystals.

Right: 'Negro's head' tourmaline with orthoclase and quartz. San Pietro in Campo, Elba (Italy)

rubellite, siberite), blue *(indicolite)*, green *(verdelite)*; there also exists a colourless tourmaline *(achroite)*, polychrome and black on the upper surface ('negro's head'); the brown variety is called *dravite*. Transparent stones are used in gem making; the lustre is vitreous. Very famous pink tourmaline is obtained at Pala near San Diego in California, also in Madagascar, Brazil, and Siberia. Other noteworthy examples of this relatively common mineral are found in the counties of Devonshire and Cornwall in England, in south western Maine, and in numerous locations in New Hampshire, Massachusetts, and Connecticut.

Dioptase

Dioptase, $CuSiO_2(OH)_2$, is a rare and much sought-after mineral, notable for its splendid emerald-green crystals. It is found at several locations in the Congo, and at Tsumeb, South West Africa. It is associated with the copper deposits of Arizona.

Inosilicates

Pyroxene group

This group contains a score or so of species, and many varieties, some of which are common constituents of metamorphic and igneous rock. They may have a granular or fibrous habit and a green or black colour. The following are among the most common examples:

Diopside, $CaMgSi_2O_6$, monoclinic; it forms short prisms or bundle-like aggregates coloured whitish, yellowish, gray, pale green, and dark green. It is sometimes transparent, with a vitreous lustre. Well known are the pale-green, transparent or translucent prismatic crystals associated with hessonite in the Val d'Ala in Piedmont. White crystals of diopside are abundant at Canaan, Connecticut, and in the neighbouring regions of Massachusetts. *Violan* is the violet magnesium variety found at San Marcel in Val d'Aosta, Italy. Good examples of green transparent chromediopside come from Outokumpu in Finland.

Hedenbergite, $CaFeSi_2O_6$, monoclinic; it is generally found in acicular, sometimes prismatic, crystals of a dark-green or gray-green colour. It is abundant in compact, fibrous-radiating masses in ore deposits at Rio Marina and Cape Calamita, and at Campiglia Marittima in Tuscany, both in Italy. Good equi-

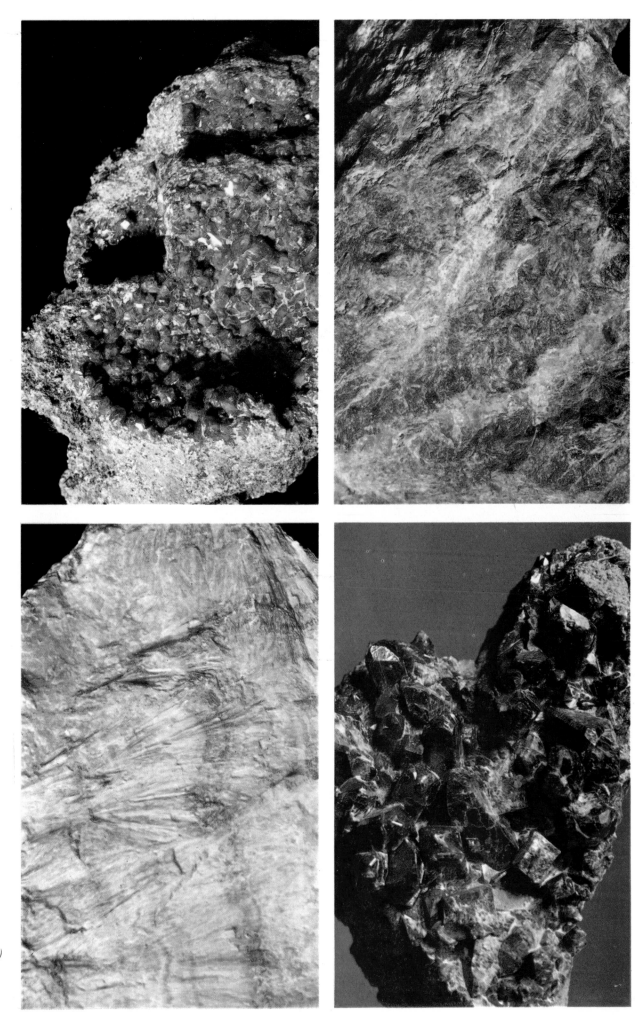

Right: Dioptase, showing its characteristic intense green colouring. Renéville (Congo)

Far right: Violan, manganiferous variety of diopside, showing the typical violet colouring. Aosta Valley (Italy)

Right: Johannsenite, slightly fibrous. Recoaro, Vicenza (Italy)

Far right: Fassaite. Traversella, Piedmont (Italy)

dimensional black crystals originate from Nordmark in Värmland, Sweden.

Johannsenite, $CaMnSi_2O_6$, monoclinic; it forms compact aggregates with a fibrous structure and a pale-green colour that darkens during oxidation; it is often associated with manganese-rich formations and occurs on Mount Civillina near Recoaro, Vicenza (Italy).

Augite has an extremely complex formula; it can contain calcium, iron, magnesium, titanium, and aluminium. The black, prismatic, and highly attractive crystals form commonly in volcanic rocks, for example around Vesuvius in Italy and Teplice in Czechoslovakia. In the United States it is obtained from Twin Peaks, Utah.

Very similar to the last mineral is *fassaite*, which has short crystals coloured gray green or dark green; it is found at Val di Fassa, Trentino, and Traversella in Piedmont (Italy).

An example of a pyroxene crystallized in the orthorhombic system is *bronzite*, $(Mg,Fe)SiO_3$, contained in various basic rocks such as gabbro; it is greenish or bronze coloured. A much rarer mineral is *carpholite*, in yellowish, radiating tufts, found in Bohemia, Harz (Germany), and Ardennes (Belgium).

Spodumene, $LiAlSi_2O_6$, is a monoclinic pyroxene which can have prismatic crystals either colourless or yellow *(triphane)*, pink *(kunzite)*, green *(hiddenite)*, transparent, or translucent. The coloured or transparent varieties are used in jewellery. The best crystals are found in Minas Gerais (Brazil), San Pedro Mine in California, and Madagascar.

Amphibole group

These are more common and more widely distributed than the pyroxenes, which at first sight they greatly resemble because they too almost always show a fibrous or acicular habit; the colour varies from white to blackish green.

Tremolite, $Ca_2Mg_5Si_8O_{22}(OH)_2$, monoclinic; it appears in fibrous aggregates, white, grayish, pale green, with a silky lustre. Good fibrous-radiating groups occur at Campolongo in the Ticino Canton (Switzerland).

Actinolite, $Ca_2(Mg,Fe)_5Si_8O_{22}(OH)_2$, monoclinic; it has the same appearance as tremolite, but it is pale green, very dark green, or blue green. It occurs in a great many places throughout the Alps, and at Franklin, New Jersey, and Mineral Hill, Pennsylvania.

Hornblende. This is a silicate with a very complex composition and can contain calcium, sodium, potassium, magnesium, iron, and aluminium. It crystallizes in the monoclinic system and can occur in fibrous and granular aggregates, or well-shaped prismatic individuals whose colour varies from green to black. Good crystal groups are found at Kragerö, Snarum, and Arendal in Norway, at Nordmark in Sweden, the volcanic formations of Lazio, Vesuvius, and Etna in

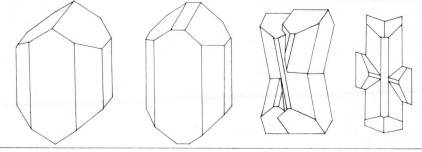

98

Right: Prismatic actinolite. St Gotthard (Italy)

Left: Kunzite, pink variety of spodumene. Pala San Diego (California)

Left: Drawings of augite: the first shows the most common and widely distributed forms; the last two are twinned crystals from Stromboli (Italy)

Italy. Black hornblende is found at various localities in Quebec and Ontario.

The very fibrous varieties of amphibole – tremolite, actinolite, or hornblende to a lesser degree – are called 'amphibole amianthus', and when the fibres are fine they are called 'byssolite'.

Glaucophane is an example of the sodium amphiboles and has dark blue acicular crystals, sometimes occurring on the surface of schist. *Crocidolite* is a very fibrous variety, white or blue, chemically similar to another sodium amphibole called *riebeckite*. It is also known as 'blue amianthus' or 'Cape amianthus', because it occurs near Cape Town in South Africa.

Wollastonite, pectolite, rhodonite, babingtonite

These triclinic inosilicates are rather similar to amphiboles, although their structure is more complex.

Wollastonite, $CaSiO_3$. It often forms at the contact between igneous intrusions and limestone, and shows fibrous masses of white, gray, or pink crystals with a silky lustre; otherwise it forms equidimensional individuals. Good individual crystals come from Natural Bridge, St Lawrence County, New York, and a compact variety from Isle Royal, Keweenaw County, Michigan; also from Riverside County, California.

Pectolite, $Ca_2NaSi_3O_8OH$. It forms fibrous-radiating groups with a white or gray-white colour and silky lustre, and it is often associated with apophyllite and zeolites such as natrolite, which has the same appearance but a vitreous instead of a silky lustre. The best examples are from the basalt caves of Prospect Park and New Street in New Jersey, and it occurs in nepheline syenite at Magnet Cove, near Hot Springs, Arkansas.

Rhodonite, $(Mn,Ca)SiO_3$. This mineral usually forms compact granular masses, only rarely prismatic or tabular crystals coloured pink, and is obtained in the famous Franklin mines in New Jersey. The compact

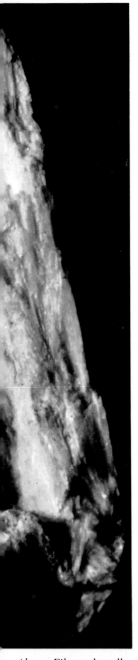

Above: Tabular crystals of rhodonite. Franklin (New Jersey)

Left: Orthoclase partially covered with brownish-black babingtonite. Baveno (Italy)

Above: Fibrous lamellar wollastonite. Jeseník, Silesia (Czechoslovakia)

Two diagrams of rhodonite (immediately below) showing an equidimensional habit on the left and a tabular habit on the right. Centre: Babingtonite from Nassau (Germany). Bottom: Babingtonite from Baveno (Italy)

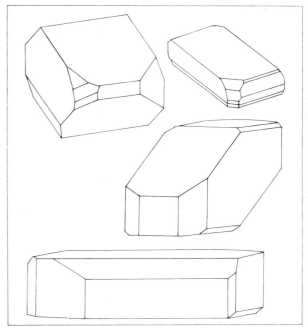

Right: Globular prehnite. Siusi Alps, Bolzano (Italy)

variety is used as an ornamental stone. It is found at Vittinge (Finland), and Vermland, Sweden.

Babingtonite, $Ca_2Fe\ddot{}\ Fe\ddot{}\ Si_5O_{14}OH$. This is not a very common mineral. It forms aggregates of small flattened violet-black crystals. At Baveno (Italy) it covers orthoclase crystals and frequently alters to limonitic products. It is also obtained from Arendal in Norway, Germany, and the United States.

Prehnite, bavenite

Prehnite, $Ca_2Al_2Si_3O_{10}(OH)_2$, does not often form separate crystals but more often minutely fibrous crusts or globular, mammillated or reniform aggregates coloured gray, yellowish, yellow green, emerald green; it is translucent with a vitreous, greasy, or pearly lustre. Prehnite in Ticino Canton (Switzerland) is a delicate green or blue. It is also found in various alpine localities such as Val d'Ala (Piedmont) and Val di Fassa (Trentino), as well as in the Apennines where it is often associated with datolite, such as Toggiano near Modena and Ciano d'Enza in Reggio Emilia (Italy). It also occurs with datolite at Paterson, New Jersey.

Bavenite, $Ca_4(Be,Al)_4Si_9(O,OH)_{28}$, forms spherical, fibrous-radiating, white or yellow aggregates, and was first found in granite at Baveno (Lake Maggiore, Italy); it also occurs in several parts of Ossola, Graubünden (Switzerland), Czechoslovakia, and California.

Phyllosilicates

Apophyllite, pyrophyllite, talc

Apophyllite, $KCa_4Si_8O_{20}(F,OH).8H_2O$, crystallizes in the tetragonal system in pseudo-cubic tabular, lamellar, and occasionally pyramidal crystals; colourless or white, blue, pink, sulphur yellow, green; transparent, translucent, or opaque. Large greenish-blue crystals with a vitreous lustre, in association with stilbite and scolecite, are mined in Rio Grande do Sul in Brazil; there are pseudo-cubic white crystals at Poona near Bombay in India; pyramidal green crystals again in India; and yellow crystals are found with harmotome in Finland. Transparent, colourless, white, or pink examples occur in the Siusi Alps near Bolzano, Italy.

Pyrophyllite, $Al_2Si_4O_{10}(OH)_2$, is orthorhombic, and typically forms aggregates consisting of plate-like individuals arranged in a fan or rosette shape, coloured silver gray, and greasy to the touch. The green, compact variety called *agalmatolite* is used for carved ornaments, on account of its attractive colour and texture.

Talc, $Mg_3Si_4O_{10}(OH)_2$, monoclinic, is more widely distributed than the previous mineral. It is found in compact masses or lamellar aggregates greasy to the touch, coloured white, gray, or green. In Val Malenco (Italy), in addition to the common white

White tabular apophyllite. Siusi Alps, Bolzano (Italy)

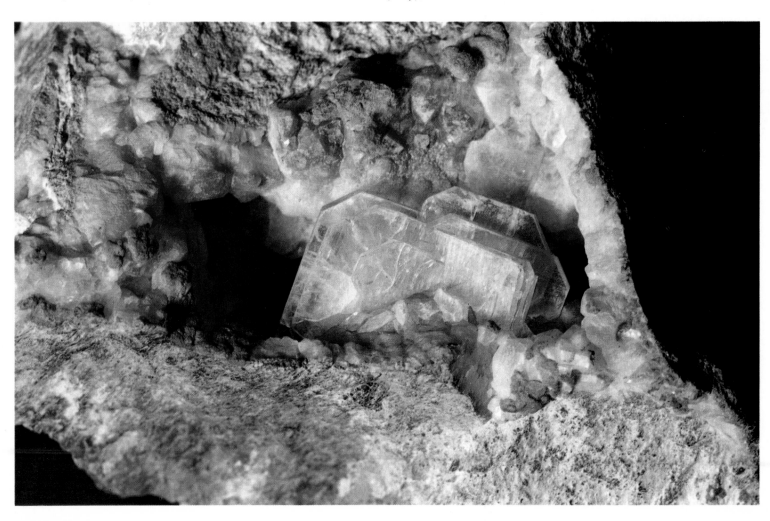

Right: Transparent prismatic apophyllite with cleavage surfaces parallel to the base. Poona, Bombay (India)

Far right: Radiating pyrophyllite plates. Urals (USSR)

Below: Bavenite. Globular plated aggregates on orthoclase. Baveno, Lake Maggiore (Italy)

103

and gray varieties, there is also green talc in micaceous plates ('precious talc').

In the United States many talc or soapstone quarries are located along the line of the Appalachian Mountains from Vermont to Georgia. The major producing states are California, North Carolina, Texas, and Georgia.

Mica group, chrysocolla

The mica group contains minerals which are almost always lamellar, with a perfect cleavage which produces very thin, transparent plates. Although they crystallize in the monoclinic system, they frequently show a hexagonal or triangular outline; they occur very frequently in igneous and metamorphic rocks. Their thermal and electrical insulating properties make them very useful in industry.

Muscovite, $KAl_3Si_3O_{10}(OH)_2$. Also called 'white mica', due to its silvery colour and pearly lustre, its crystals are composed of stacks or clusters of plates, sometimes of quite considerable size. The colour can be green due to the presence of iron or chrome, as in the *fuchsite* variety; the magnesium variety *alurgite* is a reddish colour. Quite good groups of muscovite are found in many parts of the Alps, while good crystals are obtained from Minas Gerais in Brazil and Amelia in Virginia.

Paragonite is a mica very similar to muscovite, which contains sodium, and is a clear white colour.

Biotite, $K_2(Mg,Fe^{..},Al,Fe^{...})_{4-6}(Si,Al)_8O_{20}(OH)_4$. It is known as 'black mica', due to the very dark colour of the comparatively small plates of which this highly lustrous – almost metallic – mineral is composed.

Phlogopite, on the other hand, has large hexagonal crystals; it is a mica similar to biotite but contains more magnesium and is brown or yellow brown. Single biotite crystals occur in the volcanic rock of Latium and Vesuvius (Italy). Phlogopite is found in large crystals – sometimes even 10 metres (over 30 feet) long – in Ontario, and also in New Jersey and Madagascar.

There is a lilac-coloured mica which contains lithium, called *lepidolite*, which is found in the

Above left: Lepidolite, showing its typical delicate lilac colour. Rožná, Moravia (Czechoslovakia)

Above: Alurgite, manganiferous variety of muscovite. Aosta Valley (Italy)

Concretionary mass of chrysocolla. Sa Duchessa Mine, Cagliari, Sardinia (Italy)

pegmatites of Madagascar, Brazil, Rhodesia, Australia, and Czechoslovakia. In the United States it is common in south western Maine; in California it is found in the tourmaline-bearing pegmatites of San Diego County. When found in sufficient quantities it is used for the extraction of lithium.

Chrysocolla, $CuSiO_3.2H_2O$, forms stalactitic masses or amorphous crusts, green, blue, sky blue, with a porcellaneous or vitreous lustre. This secondary mineral is very common in copper deposits. Good examples are obtained from the Clifton-Morenci district, Greenlee County, and from Gila and Cochise Counties, all in Arizona.

Chlorite group

The chlorites look similar to the micas. They are lamellar, dark green or blue green, and are given different names according to their composition and structure, although it is difficult to identify them at first sight: *penninite, clinochlore, ripidolite,* etc.

Above: Lamellar chlorite. Achmatovsk, Urals (USSR)

Above right: Vein of chrysocolla (or serpentine asbestos), in altered serpentinite. Silesia (Poland)

Right: Garnierite, a variety of chrysocolla rich in nickel, used for the extraction of this metal. Nouméa (New Caledonia)

Kämmererite is a chromiferous chlorite of a dark-violet or violet-red colour, found at Lake Itkul near Sverd-lovsk, USSR. It is abundant in Lancaster County, Pennsylvania, while the variety penninite occurs at Snake Creek, Wasatch County, Utah. Clinochlore is probably the most common of the chlorite minerals, and it forms large plates in talc at West Chester, Chester County, Pennsylvania.

Serpentine, $Mg_3Si_2O_5(OH)_4$, is mentioned here only because it is an essential constituent of the masses of green rocks known as 'serpentinites', which are fairly common in the Alps and the Apen-nines and are associated with several interesting minerals such as demantoid, magnetite, pyrite, perovskite, artinite, etc. It comprises a lamellar variety called *antigorite*, and a fibrous one called *chrysotile* which constitutes the well-known 'asbestos serpentine'. The colour of chrysotile can be white, yellow, or greenish, with a silky lustre and exception-ally long fibres: those at Campo Franscia in Val Malenco (Lombardy), are more than a metre in length. Serpentine in veins, coatings, and nodules of an attractive apple-green, yellow, and green colour are found in metamorphosed magnesium rocks; if it can be polished it is called 'precious serpentine'. Extensive deposits of serpentine occur in Quebec.

Tektosilicates

Feldspathoid group
These minerals are grouped together for petro-graphic reasons, not systematic ones. They are

associated with igneous rocks that are poor in silicon.

Nepheline, $(Na,K)AlSiO_4$, forms short, prismatic, hexagonal or sometimes tabular crystals, whitish, gray, or blue, with a vitreous or greasy lustre. Fine specimens occur in lava from Vesuvius and from Latium; the compact, reddish or gray variety is called *eleolite* and is found in Scandinavian countries.

Analcite, (analcime), $NaAlSi_2O_6.H_2O$: isometric system; sg 2.22; hardness 6. Its crystals, which are sometimes very large, have a tetragonal tristoctahedral habit and are white, pale pink, and pale gray green; transparent, translucent, or opaque, with a vitreous lustre. Large pink crystals come from the Siusi Alps, Italy, and good specimens are abundant in the copper district in Keeweenaw County, Michigan.

Leucite, $KAlSi_2O_6$, the name of which derives from the Greek word *'leukòs'* meaning white, has a specific gravity of 2.45 and a hardness of 5.5.

Its crystal habit is icositetrahedral *(leucitohedra)*, similar to analcite, but its colour is white or grayish,

with an earthy opaqueness. It is a typical mineral of tuff in the Italian region of Latium (Parco Chigi in Ariccia), also the lava of Roccamonfina (Campania) and Vesuvius; it is not unusual to find perfect crystals together with melanite, augite, etc.

The feldspars

These are the most commonly found silicates, especially in igneous and metamorphic rocks where crystals can be found without much difficulty.

Orthoclase, $KAlSi_3O_8$: monoclinic system; sg 2.56; hardness 6. It forms irregular granules, spathic masses and prismatic, tabular, or equidimensional crystals. It may be colourless, white, gray, pink, or red; and transparent, translucent, or opaque.

Adularia is a white or transparent orthoclase, often covered with green chlorite, typical of Alpine crystalline schist – it is in fact named after the Adula group in Graubünden, Switzerland. Sometimes an excessive development of one of the faces can cause the

mineral to have a pseudo-rhombohedral appearance.

Sanidine is an orthoclase with a tabular, often twinned, habit; coloured gray, brown, reddish, etc, it is found mainly in volcanic rock as at Soriano near Viterbo (Italy), the Island of Pantelleria, Germany, Czechoslovakia, etc. Orthoclase is common in the pegmatites of the New England states and occurs at Rossie near Grasse Lake and in the town of Hammond, both in St Lawrence County, New York. Fine twinned crystals are found near Karlovy Vary in Czechoslovakia, and large well-formed crystals can be found in some of the granite masses of Cornwall, England.

Microcline has the same chemical composition as orthoclase, but it crystallizes in the triclinic system. A leek-green variety is called *amazonite*, forms compact masses or groups of equidimensional crystals, and is obtained from Pikes Peak in Colorado, Amelia in Virginia, Brazil, and Madagascar.

The *plagioclase* series is a triclinic group of feldspars beginning with a sodium silicate called *albite*, ($NaAlSi_3O_8$) and going through the minerals with an intermediate composition to end with *anorthite*, $CaAl_2Si_2O_8$, a calcium silicate.

They are very abundant in rocks, especially igneous ones, which are classified according to the type of plagioclase present. Albite forms aggregates of white, flattened crystals which can sometimes be transparent, with a vitreous lustre; long white crystals, often covered with chlorite, sometimes twinned, belong to the *pericline* variety, and are quite widely distributed in Alto Adige (Italy), and in Austria. Albite is found at numerous localities in the United States, among which are the mica mines near Amelia Court House, Amelia County, Virginia, and the Pikes Peak area of Teller County, Colorado.

Anorthite crystals are found in the lavas of the island of Miyake, Japan, and are obtained from Franklin, Sussex County, New Jersey.

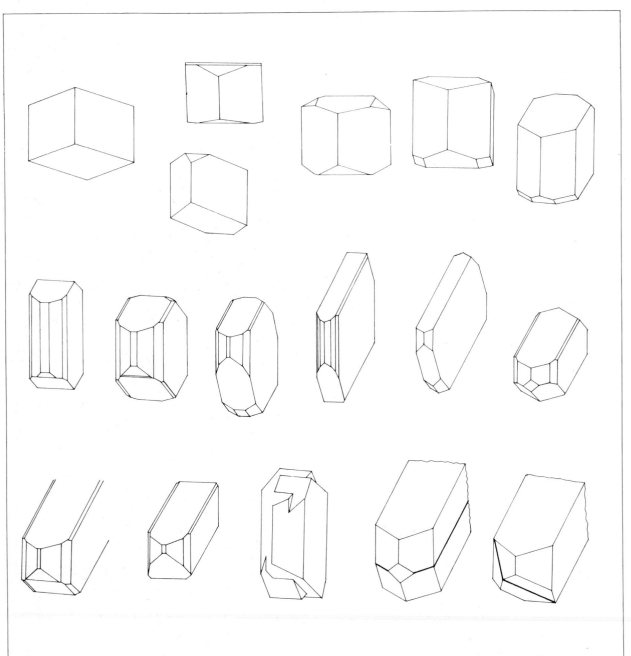

Potassium feldspar. The first four in the top row are typical forms of adularia; the others show normal forms of orthoclase of granitic and pegmatitic rocks and also the flattened forms (sanidine) of volcanic rocks; the last three drawings show twinned crystals

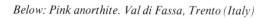

Above: Albite; simple and polysynthetic twinning

Below left: Sanidine twinned on the Carlsbad law. Enfola (Elba)

Below: Pink anorthite. Val di Fassa, Trento (Italy)

Above: A crystal of the amazonstone variety of microcline. Pikes Peak, Colorado

Danburite, sodalite, haüynite, lazurite, scapolite

Danburite, $CaB_2Si_2O_8$, is orthorhombic, very similar in habit to topaz, and shows prismatic crystals with wedge-shaped ends, which are transparent or translucent. It can be colourless, white, pink, or yellow. Good groups of crystals occur at San Luis Potosí in Mexico, and it has been found in Graubünden, Switzerland. It is found in fine crystals at Russell, St Lawrence County, New York.

Sodalite is a feldspathoid which only occasionally forms separate crystals, more frequently compact granular masses of a blue colour which are used for ornamental purposes, from Canada and Brazil.

Haüynite forms small green-gray crystals with a vitreous lustre, or blue granules in volcanic rock as for example at Ariccia near Rome, and Vulture in Basilicata, Italy, and in Tahiti.

Lazurite, better known as *lapis lazuli*, is always in blue or dark-blue granular masses and is much used as an ornamental stone. There are very famous deposits at Badakhshan in Afghanistan, Lake Baikal in Siberia, and also in Chile.

Scapolite or *wernerite*, is a mixture of two complex silicates, one containing sodium *(marialite)*, the other calcium *(meionite)*; the intermediate varieties are called *dipyre* and *mizzonite*. It crystallizes in the tetragonal system in colourless, white, yellow, grayish, or pale-green prismatic individuals which can sometimes be transparent. It is common in limestones close to contacts with magmatic rocks, volcanic rock, and Alpine schist. Hyaline crystals are found in Ticino Canton, Switzerland; a transparent yellow variety from Brazil and Madagascar is used in gem making. Wernerite is found at Boston, Massachusetts, at numerous localities in New York State, and at Franklin, New Jersey.

The zeolites

The minerals in this group, characterized by the ease with which they undergo swelling under a flame, are very common in cavities of volcanic rock in association with datolite, prehnite, apophyllite, and many others. About twenty-five species are known, the most common of which are described below.

Above: Pericline variety of albite. Valle Aurina, Bolzano (Italy)

Right: Prismatic danburite. Charcas, San Luis Potosí (Mexico)

Natrolite, $Na_2Al_2Si_3O_{10}.2H_2O$, orthorhombic; it usually forms fibrous radiating aggregates with a vitreous lustre and a white or pale-pink colour. Well-terminated crystals are found at Livingstone, Montana; San Benito, California; the Rhineland, Hessen, Saxony (Germany); in Iceland; at Puy de Dôme in France; also near Teplice, Czechoslovakia; near Brevik, Norway; at Bishopton, Renfrew, Scotland; and near Belfast, Northern Ireland.

Scolecite, $CaAl_2Si_3O_{10}.3H_2O$, monoclinic; it is difficult to distinguish it from natrolite, except for the fact that when exposed to a flame it melts into a contorted worm-like shape (which in Greek is called *scolex*). Fine examples are found in India, Brazil, and Iceland.

The white, reddish, gray-green fibrous crystals of *thomsonite*, a sodium and calcium zeolite, are found in New Jersey and Minnesota; in the Maderaner Tal,

Right: Lapis lazuli or lazurite, smooth and compact. Afghanistan

Left: Haüynite. Ariccia, near Rome (Italy)

*Natrolite, fibrous-
radiating. Siusi Alps,
Bolzano (Italy)*

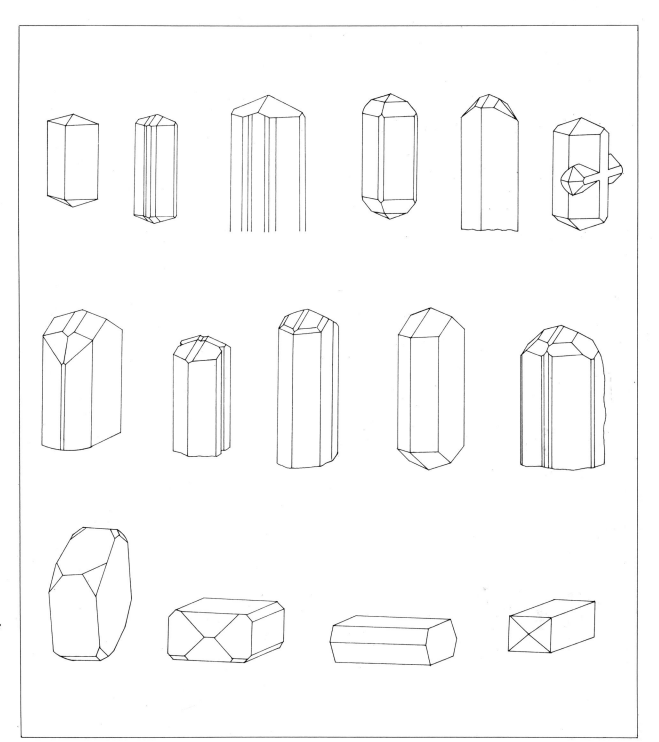

Top row: Prismatic natrolite. Middle row: Prismatic scolecite; the second from the left is twinned. Bottom row: Some typical forms of heulandite; the first is from Scotland, the second and third from Italy, the fourth from Sweden

Uri, Switzerland; and in Iceland, at Theigarhorn on the Berufjord on the east coast.

Mesolite is also a sodium and calcium zeolite, forming needle-like or filiform white crystals arranged in radiating groups or felted surfaces, which are sometimes compact. It is easily confused with the other fibrous zeolites. In the United States it occurs with other zeolites on Fritz Island in the Schuylkill River, Berks County, Pennsylvania.

The name mesolite should not be confused with *mesotype*, a term much used especially in the last century to describe unidentified fibrous zeolites which have since shown themselves to be natrolite, scolecite, or mesolite.

Dachiardite is· a rare monoclinic zeolite containing potassium, sodium, and calcium. It forms small whitish crystals, often twinned, in the pegmatites of San Piero in Campo, Elba. In these same veins there occurs *foresite*, first identified as a pure zeolite and later seen to be a mixture of stilbite, lithium mica, lithium chloride, and quartz.

Laumontite, $CaAl_2Si_4O_{12}.4H_2O$, monoclinic; this is a zeolite which is very susceptible to dehydration and crumbling away, which is why it is preserved in containers filled with water. It has striated, white, prismatic crystals, and occurs in the Alpine gneiss of Ticino Canton, Switzerland, and Beura in Ossola, Italy, and also in granite at Baveno (Lake Maggiore).

A brick-red, radiating, finely fibrous variety of *mordenite* is called *arduinite*.

Above: Thomsonite in lamellar aggregates. Siusi Alps, Bolzano (Italy)

Left: Radiating scolecite. Rio Grande do Sul (Brazil)

Below: Arduinite, a variety of mordenite. Val dei Zuccanti, Vicenza (Italy)

121

Ferrierite is an orthorhombic zeolite of sodium and potassium, which forms minute lamellar, radiating, vitreous aggregates coloured gray or brick red. It is found in very few parts of the world, among which is a railroad cutting near Kamloops Lake, British Columbia, Canada.

Heulandite, , $(Na,Ca)_{4-6}Al_6(Al,Si)_4Si_{26}O_{72}.24H_2O$, monoclinic; its crystals are mainly flattened, coloured white, brown, brick red, with a pearly lustre. White crystals come from Theigarhorn in Iceland, brown ones from Rio Grande do Sul, Brazil. In the United States it is found in the trap rocks of north eastern New Jersey at Bergen Hill, West Paterson, and Great Notch, and in minute crystals at Jones's Falls, near Baltimore, Maryland.

Stilbite, or *desmine*, $NaCa_2Al_5Si_{13}O_{36}.14H_2O$, monoclinic; it is generally found in aggregates of divergent intergrown crystals in a sheaf formation, coloured white, brown, pink, reddish, with a pearly lustre particularly on the cleavage surface. The white examples from Theigarhorn in Iceland are much sought-after, as are the yellowish-brown ones from New Jersey, and the brown radiating fan-shaped crystals from Rio Grande do Sul in Brazil associated with apophyllite, scolecite, heulandite, etc. In Poona, India, it is found in association with apophyllite. It is also found at Giebelbach near Fiesch in the Rhône Valley, Valais, Switzerland.

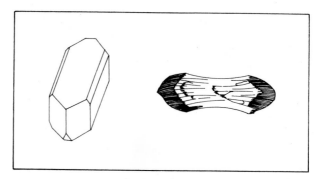

Left: Two forms of stilbite. The second is the 'sheaf' variety

Above: Radiating lamellar aggregates of stilbite. Rio Grande do Sul (Brazil)

Right: Crystals of heulandite. Val di Fassa, Trento (Italy)

Far left: Heulandite. Val di Fassa, Trento (Italy)

Left: Stilbite sheaf. Faeroe Islands

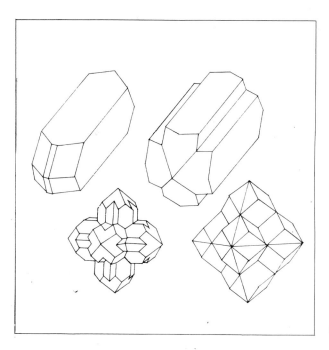

Phillipsite

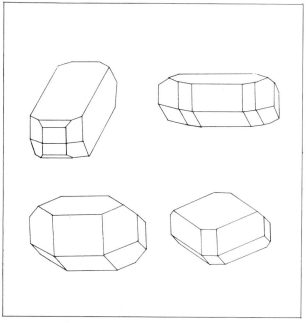

Harmotome

The *puflerite* **variety** is so-called after the Puflerloch or Pufelserloch locality (Forra di Bulla) in the Siusi Alps near Bolzano in Italy, where it forms in crusts of small, vitreous, gray globules.

Spherostilbite is simply a fibrous-radiating thomsonite. *Epistilbite*, however, is a zeolite, with the same composition as stilbite but with a lower water content; it is much rarer. It forms small reddish crystals at Margaretsville, Nova Scotia, Canada.

Phillipsite, $(K_2,Na_2,Ca)Al_2Si_4O_{12}.4\frac{1}{2}H_2O$, monoclinic; it has small, many-faced, vitreous-white crystals, often twinned in a cross formation, or it has a greater symmetry (tetragonal or isometric). This zeolite is widely distributed, though it only occurs in small quantities and its crystal groups are not very spectacular.

Gismondine, $CaAl_2Si_2O_8.4H_2O$, is monoclinic and pseudo-orthorhombic, and forms small, vitreous, white or gray crystals which are often associated with phil-

Stilbite crystals. Poona, Bombay (India)

125

Left: Harmotome in twinned crystals. Kornsnäs (Finland)

lipsite, especially in the volcanic leucite rocks of Latium in Italy.

Harmotome, $BaAl_2Si_6O_{16}.6H_2O$, monoclinic; it is rather similar to the previous mineral, but its white crystals twinned in a cross formation are generally larger. The best examples come from Strontian and Dumbarton in Scotland. It is found as small brown crystals with stilbite on the gneiss of New York Island and near Port Arthur, Lake Superior.

Chabazite, $CaAl_2Si_4O_{12}.6H_2O$, trigonal; it is also called trigonal zeolite because the rhombohedral crystals have a pseudo-cubic appearance; there is either no colour at all or it is pale yellow, brown, or flesh pink, with a vitreous lustre. Large crystals are obtained at Wasson's Bluff in Nova Scotia, and Styria in Austria. In the United States chabazite is found at Goble, Columbia County, Oregon, in the trap rocks at West Paterson and other localities in north eastern New Jersey, and at Jones's Falls, near Baltimore, Maryland.

Right: Pseudo-cubic individuals of white chabazite with pink apophyllite. Siusi Alps, Bolzano (Italy)

Bibliography

Agricola, G., *De Natura Fossilium* (1546), translated from first Latin edition by M. C. and J. A. Bandy, Geological Society of America Special Paper, New York 1955.

Bateman, A. M., *Economic mineral deposits*, second edition, John Wiley and Sons, New York 1950.

Bishop, A. C., *An outline of crystal morphology*, London 1967.

Bragg, W. L., and Claringbull, G. F., *The crystal structure of minerals,* G. Bell and Sons, London 1965.

Correns, C. W., *Introduction to mineralogy, crystallography and petrology,* translated by W. D. Johns, George Allen and Unwin, London 1969.

Dana, E. S., *A textbook of mineralogy*, fourth edition revised and enlarged by W. E. Ford, John Wiley and Sons, New York 1949.

Dana, J. D., *A system of mineralogy*, seventh edition, John Wiley and Sons, New York, vol. I, 1944; vol. II, 1951; vol. III, 1962.

Deer, W. A., Howie, R. A., and Zussman, J., *An introduction to the rock-forming minerals,* Longmans, London 1966.

Desautels, P. E., *The mineral kingdom*, Hamlyn, Feltham 1969.

Hey, M. H., *Chemical index of minerals,* second edition, London 1955 (appendix 1963).

Hurlbut, C. S., *Minerals and man*, Thames and Hudson, London 1969.

Jones, W. R., *Minerals in industry*, Penguin Books, Harmondsworth 1963.

Kostov, I., *Mineralogy*, Edinburgh and London 1968.

Kraus, E. J., Hunt, W. F. and Ramsdell, L. S., *Mineralogy: an introduction to the study of minerals and crystals,* fifth edition, McGraw-Hill, New York 1959.

Phillips, F. C., *An introduction to crystallography*, Longmans, London and New York 1964.

Pough, F. H., *A field guide to rocks and minerals*, Cambridge (Massachusetts) 1960.

Read, H. H., *Rutley's elements of mineralogy*, twenty-sixth edition, Thomas Murby and Company, London 1970.

Sinkankas, J., *Gemstones of North America*, D. Van Nostrand, Princeton 1959.

Tunell, G. and Murdoch, J., *Introduction to crystallography*, W. H. Freeman and Company, San Francisco 1959.

U.S. Bureau of Mines, *Minerals yearbook*, Washington, D.C.

Voskuil, W. H., *Minerals in world industry*, McGraw-Hill, New York 1955.

Werner, A. G., *On the external characters of minerals* (1774), translated by A. V. Carozzi, University of Illinois Press, Urbana 1962.

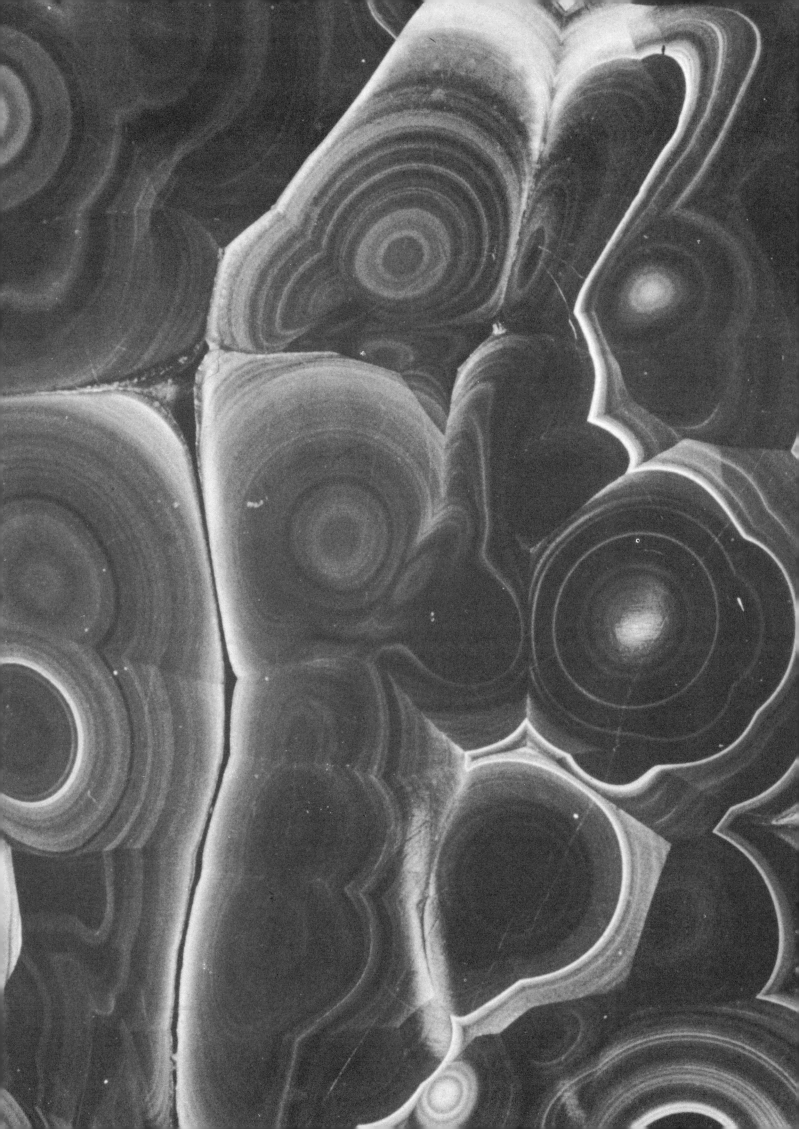